ASE

Medium/heavy duty
truck technician
certification series.

D0764904

Medium/Heavy Duty Truck Certification Series

Electrical/Electronic Systems (T6)
5th Edition

DELMAR
CENGAGE Learning™

Australia • Brazil • Japan • Korea • Mexico • Singapore • Spain • United Kingdom • United States

ASE Test Preparation: Medium/Heavy Duty Truck Certification Series, Electrical/Electronic Systems (T6), Fifth Edition

Vice President, Technology and Trades
   Professional Business Unit:
   Gregory L. Clayton

Director, Professional Transportation Industry
   Training Solutions: Kristen L. Davis

Editorial Assistant: Danielle Filippone

Director of Marketing: Beth A. Lutz

Marketing Manager: Jennifer Barbic

Senior Production Director: Wendy Troeger

Production Manager: Sherondra Thedford

Content Project Management: PreMediaGlobal

Senior Art Director: Benjamin Gleeksman

Section Opener Image: Image Copyright Marek
   Pawluczuk, 2012. Used under license from
   Shutterstock.com

For product information and technology assistance, contact us at
Cengage Learning Customer & Sales Support, 1-800-354-9706

For permission to use material from this text or product,
submit all requests online at **www.cengage.com/permissions**
Further permissions questions can be emailed to
permissionrequest@cengage.com

ISBN-13: 978-1-111-12902-6

ISBN-10: 1-111-12902-9

Delmar
Executive Woods
5 Maxwell Drive
Clifton Park, NY 12065
USA

Cengage Learning is a leading provider of customized learning solutions with office locations around the globe, including Singapore, the United Kingdom, Australia, Mexico, Brazil, and Japan. Locate your local office at **www.cengage.com/global**

Cengage Learning products are represented in Canada by Nelson Education, Ltd.

To learn more about Delmar, visit **www.cengage.com/delmar**

Purchase any of our products at your local bookstore or at our preferred online store **www.cengagebrain.com**

Notice to the Reader

Publisher does not warrant or guarantee any of the products described herein or perform any independent analysis in connection with any of the product information contained herein. Publisher does not assume, and expressly disclaims, any obligation to obtain and include information other than that provided to it by the manufacturer. The reader is expressly warned to consider and adopt all safety precautions that might be indicated by the activities described herein and to avoid all potential hazards. By following the instructions contained herein, the reader willingly assumes all risks in connection with such instructions. The publisher makes no representations or warranties of any kind, including but not limited to, the warranties of fitness for particular purpose or merchantability, nor are any such representations implied with respect to the material set forth herein, and the publisher takes no responsibility with respect to such material. The publisher shall not be liable for any special, consequential, or exemplary damages resulting, in whole or part, from the readers' use of, or reliance upon, this material.

Printed in the United States of America
2 3 4 5 6   18 17 16 15 14

Table of Contents

SECTION 6 **Answer Keys and Explanations 107**

SECTION 7 **Appendices 233**

Delmar, a part of Cengage Learning, is very pleased that you have chosen to use our ASE Test Preparation Guide to help prepare yourself for the Electrical/Electronic Systems (T6) ASE certification examination. This guide is designed to help prepare you for your actual exam by providing you with an overview and introduction of the testing process, introducing you to the task list for the Electrical/Electronic Systems (T6) certification exam, giving you an understanding of what knowledge and skills you are expected to have in order to successfully perform the duties associated with each task area, and providing you with several preparation exams designed to emulate the live exam content in hopes of assessing your overall exam readiness.

If you have a basic working knowledge of the discipline you are testing for, you will find this book is an excellent guide, helping you understand the "must know" items needed to successfully pass the ASE certification exam. This manual is not a textbook. Its objective is to prepare the individual who has the existing requisite experience and knowledge to attempt the challenge of the ASE certification process. This guide cannot replace the hands-on experience and theoretical knowledge required by ASE to master the vehicle repair technology associated with this exam. If you are unable to understand more than a few of the preparation questions and their corresponding explanations in this book, it could be that you require either more shop-floor experience or further study.

This book begins by providing an overview of, and introduction to, the testing process. This section outlines what we recommend you do to prepare, what to expect on the actual test day, and overall methodologies for your success. This section is followed by a detailed overview of the ASE task list to include explanations of the knowledge and skills you must possess to successfully answer questions related to each particular task. After the task list, we provide six sample preparation exams for you to use as a means of evaluating areas of understanding, as well as areas requiring improvement in order to successfully pass the ASE exam. Delmar is the first and only test preparation organization to provide so many unique preparation exams. We enhanced our guides to include this support as a means of providing you with the best preparation product available. Section 6 of this guide includes the answer keys for each preparation exam, along with the answer explanations for each question. Each answer explanation also contains a reference back to the related task or tasks that it assesses. This will provide you with a quick and easy method for referring back to the task list whenever needed. The last section of this book contains blank answer sheet forms you can use as you attempt each preparation exam, along with a glossary of terms.

OUR COMMITMENT TO EXCELLENCE

Thank you for choosing Delmar, Cengage Learning for your ASE test preparation needs. All of the writers, editors, and Delmar staff have worked very hard to make this test preparation guide second to none. We feel confident that you will find this guide easy to use and extremely beneficial as you prepare for your actual ASE exam.

Delmar, Cengage Learning has sought out the best subject matter experts in the country to help with the development of *ASE Test Preparation: Medium/Heavy Duty Truck Certification Series, Electrical/Electronic Systems (T6), 5th Edition*. Preparation questions are authored and then reviewed by a group of certified, subject-matter experts to ensure the highest level of quality and validity to our product.

If you have any questions concerning this guide or any guide in this series, please visit us on the web at **http://www.trainingbay.cengage.com**.

For web-based online test preparation for ASE certifications, please visit us on the web at **http://www.techniciantestprep.com/** to learn more.

ABOUT THE AUTHOR

Jerry Clemons has been around cars, trucks, equipment, and machinery throughout his whole life. Being raised on a large farm in central Kentucky provided him with an opportunity to complete mechanical repair procedures from an early age. Jerry earned an associate in applied science degree in Automotive Technology from Southern Illinois University and a bachelor of science degree in Vocational, Industrial, and Technical Education from Western Kentucky University. Jerry has also completed a master of science degree in Safety, Security, and Emergency Management from Eastern Kentucky University. Jerry has been employed at Elizabethtown Community and Technical College since 1999 and is currently an associate professor for the Automotive and Diesel Technology Programs. Jerry holds the following ASE certifications: Master Medium/Heavy Truck Technician, Master Automotive Technician, Advanced Engine Performance (L1), Truck Equipment Electrical Installation (E2), and Automotive Service Consultant (C1). Jerry is a member of the Mobile Air Conditioning Society (MACS) as well as a member of the North American Council of Automotive Teachers (NACAT). Jerry has been involved in developing transportation material for Cengage Learning for seven years.

ABOUT THE SERIES ADVISOR

Brian (BJ) Crowley has experienced several different aspects of the diesel industry over the past 10 years. Now a diesel technician in the oil and gas industry, BJ owned and operated a diesel repair shop where he repaired heavy, medium, and light trucks, as well as agricultural and construction equipment. He earned an associate's degree in diesel technology from Elizabethtown Community and Technical College in Kentucky and is an ASE Master certified medium/heavy truck technician.

The History and Purpose of ASE

ASE began as the National Institute for Automotive Service Excellence (NIASE). It was founded as a nonprofit, independent entity in 1972 by a group of industry leaders with the single goal of providing a means for consumers to distinguish between incompetent and competent technicians. It accomplishes this goal through the testing and certification of repair and service professionals. Though it is still known as the National Institute for Automotive Service Excellence, it is now called "ASE" for short.

Today, ASE offers more than 40 certification exams in automotive, medium/heavy duty truck, collision repair and refinish, school bus, transit bus, parts specialist, automobile service consultant, and other industry-related areas. At this time there are more than 385,000 professionals nationwide with current ASE certifications. These professionals are employed by new car and truck dealerships, independent repair facilities, fleets, service stations, franchised service facilities, and more.

ASE's certification exams are industry-driven and cover practically every on-highway vehicle service segment. The exams are designed to stress the knowledge of job-related skills. Certification consists of passing at least one exam and documenting two years of relevant work experience. To maintain certification, those with ASE credentials must be re-tested every five years.

While ASE certifications are a targeted means of acknowledging the skills and abilities of an individual technician, ASE also has a program designed to provide recognition for highly qualified repair, support, and parts businesses. The Blue Seal of Excellence Recognition Program allows businesses to showcase their technicians and their commitment to excellence. One of the requirements of becoming Blue Seal recognized is that the facility must have a minimum of 75 percent of its technicians ASE certified. Additional criteria apply, and program details can be found on the ASE website.

ASE recognized that educational programs serving the service and repair industry also needed a way to be recognized as having the faculty, facilities, and equipment necessary to provide a quality education to students wanting to become service professionals. Through the combined efforts of ASE, industry, and education leaders, the nonprofit organization entitled the National Automotive Technicians Education Foundation (NATEF) was created in 1983 to evaluate and recognize academic programs. Today more than 2,000 educational programs are NATEF certified.

For additional information about ASE, NATEF, or any of their programs, the following contact information can be used:

>National Institute for Automotive Service Excellence (ASE)
>101 Blue Seal Drive S.E.
>Suite 101
>Leesburg, VA 20175
>Telephone: 703-669-6600
>Fax: 703-669-6123
>Website: **www.ase.com**

Overview and Introduction

Participating in the National Institute for Automotive Service Excellence (ASE) voluntary certification program provides you with the opportunity to demonstrate you are a qualified and skilled professional technician who has the "know-how" required to successfully work on today's modern vehicles.

EXAM ADMINISTRATION

Note: After November 2011, ASE will no longer offer paper and pencil certification exams. There will be no Winter testing window in 2012, and ASE will offer and support CBT testing exclusively starting in April 2012.

ASE provides computer-based testing (CBT) exams, which are administered at test centers across the nation. It is recommended that you go to the ASE website at *http://www.ase.com* and review the conditions and requirements for this type of exam. There is also an exam demonstration page that allows you to personally experience how this type of exam operates before you register.

CBT exams are available four times annually, for two-month windows, with a month of no testing in between each testing window:

- January/February – Winter testing window
- April/May – Spring testing window
- July/August – Summer testing window
- October/November – Fall testing window

Please note, testing windows and timing may change. It is recommended you go to the ASE website at *http://www.ase.com* and review the latest testing schedules.

UNDERSTANDING TEST QUESTION BASICS

ASE exam questions are written by service industry experts. Each question on an exam is created during an ASE-hosted "item-writing" workshop. During these workshops, expert service representatives from manufacturers (domestic and import), aftermarket parts and equipment manufacturers, working technicians, and technical educators gather to share ideas and convert them into actual exam questions. Each exam question written by these experts must then survive review by all members of the group. The questions are designed to address the practical application of repair and diagnosis knowledge and skills practiced by technicians in their day-to-day work.

After the item-writing workshop, all questions are pre-tested and quality-checked on a national sample of technicians. Those questions that meet ASE standards of quality and accuracy are included in the scored sections of the exams; the "rejects" are sent back to the drawing board or discarded altogether.

Depending on the topic of the certification exam, you will be asked between 40 and 80 multiple-choice questions. You can determine the approximate number of questions you can expect to be asked during the Electrical/Electronic Systems (T6) certification exam by reviewing the task list in Section 4 of this book. The five-year recertification exam will cover this same content; however, the number of questions for each content area of the recertification exam will be reduced by approximately one-half.

> *Note:* Exams may contain questions that are included for statistical research purposes only. Your answers to these questions will not affect your score, but since you do not know which ones they are, you should answer all questions in the exam.

Using multiple criteria, including cross-sections by age, race, and other background information, ASE is able to guarantee that exam questions do not include bias for or against any particular group. A question that shows bias toward any particular group is discarded.

TEST-TAKING STRATEGIES

Before beginning your exam, quickly look over the exam to determine the total number of questions that you will need to answer. Having this knowledge will help you manage your time throughout the exam to ensure you have enough available to answer all of the questions presented. Read through each question completely before marking your answer. Answer the questions in the order they appear on the exam. Leave the questions blank that you are not sure of and move on to the next question. You can return to those unanswered questions after you have finished the others. These questions may actually be easier to answer at a later time once your mind has had additional time to consider them on a subconscious level. In addition, you might find information in other questions that will help you recall the answers to some of them.

Multiple-choice exams are sometimes challenging because there are often several choices that may seem possible, or partially correct, and therefore it may be difficult to decide on the most appropriate answer choice. The best strategy, in this case, is to first determine the correct answer before looking at the answer options. If you see the answer you decided on, you should still be careful to examine the other answer options to make sure that none seems more correct than yours. If you do not know or are not sure of the answer, read each option very carefully and try to eliminate those options that you know are incorrect. That way, you can often arrive at the correct choice through a process of elimination.

If you have gone through the entire exam, and you still do not know the answer to some of the questions, *then guess*. Yes, guess. You then have at least a 25 percent chance of being correct. While your score is based on the number of questions answered correctly, any question left blank, or unanswered, is automatically scored as incorrect.

There is a lot of "folk" wisdom on the subject of test taking that you may hear about as you prepare for your ASE exam. For example, there are those who would advise you to avoid response options that use certain words such as *all, none, always, never, must,* and *only,* to name a few. This, they claim, is because nothing in life is exclusive. They would advise you to choose response options that use words that allow for some exception, such as *sometimes, frequently, rarely, often, usually, seldom,* and *normally.* They would also advise you to avoid the first and last option (A or D) because exam writers, they feel, are more comfortable if they put the correct answer in the middle (B or C) of the choices. Another recommendation often offered is to select the option that is either shorter or longer than the other three choices because it is more likely to be correct. Some would advise you to never change an answer since your first intuition is usually correct. Another area of "folk" wisdom focuses specifically on any repetitive patterns created by your question responses (e.g., A, B, C, A, B, C, A, B, C).

Many individuals may say that there are actual grains of truth in this "folk" wisdom, and whereas with some exams, this may prove true, it is not relevant in regard to the ASE certification exams. ASE validates all exam questions and test forms through a national sample of technicians, and only those questions and test forms that meet ASE standards of quality and accuracy are included in the scored sections of the exams. Any biased questions or patterns are discarded altogether, and therefore, it is highly unlikely you will experience any of this "folk" wisdom on an actual ASE exam.

PREPARING FOR THE EXAM

Delmar, Cengage Learning wants to make sure we are providing you with the most thorough preparation guide possible. To demonstrate this, we have included hundreds of preparation questions in this guide. These questions are designed to provide as many opportunities as possible to prepare you to successfully pass your ASE exam. The preparation approach we recommend and outline in this book is designed to help you build confidence in demonstrating what task area content you already know well while also outlining what areas you should review in more detail prior to the actual exam.

We recommend that your first step in the preparation process should be to thoroughly review Section 3 of this book. This section contains a description and explanation of the type of questions you will find on an ASE exam.

Once you understand how the questions will be presented, we then recommend that you thoroughly review Section 4 of this book. This section contains information that will help you establish an understanding of what the exam will be evaluating, and specifically, how many questions to expect in each specific task area.

As your third preparatory step, we recommend you complete your first preparation exam, located in Section 5 of this book. Answer one question at a time. After you answer each question, review the answer and question explanation information located in Section 6. This section will provide you with instant response feedback, allowing you to gauge your progress, one question at a time, throughout this first preparation exam. If after reading the question explanation you do not feel you understand the reasoning for the correct answer, go back and review the task list overview (Section 4) for the task that is related to that question. Included with each question explanation is a clear identifier of the task area that is being assessed (e.g., Task A.1). If at that point you still do not feel you have a solid understanding of the material, identify a good source of information on the topic, such as an educational course, textbook, or other related source of topical learning, and do some additional studying.

After you have completed your first preparation exam and have reviewed your answers, you are ready to complete your next preparation exam. A total of six practice exams are available in Section 5 of this book. For your second preparation exam, we recommend that you answer the questions as if you were taking the actual exam. Do not use any reference material or allow any interruptions in order to get a feel for how you will do on the actual exam. Once you have answered all of the questions, grade your results using the Answer Key in Section 6. For every question that you gave an incorrect answer to, study the explanations to the answers and/or the overview of the related task areas. Try to determine the root cause for missing the question. The easiest thing to correct is learning the correct technical content. The hardest things to correct are behaviors that lead you to an incorrect conclusion. If you knew the information but still got the question incorrect, there is likely a test-taking behavior that will need to be corrected. An example of this would be reading too quickly and skipping over words that affect your reasoning. If you can identify what you did that caused you to answer the question incorrectly, you can eliminate that cause and improve your score.

Here are some basic guidelines to follow while preparing for the exam:

- Focus your studies on those areas you are weak in.
- Be honest with yourself when determining if you understand something.
- Study often but for short periods of time.
- Remove yourself from all distractions when studying.
- Keep in mind that the goal of studying is not just to pass the exam; the real goal is to learn.
- Prepare physically by getting a good night's rest before the exam, and eat meals that provide energy but do not cause discomfort.
- Arrive early to the exam site to avoid long waits as test candidates check in.
- Use all of the time available for your exams. If you finish early, spend the remaining time reviewing your answers.
- Do not leave any questions unanswered. If absolutely necessary, guess. All unanswered questions are automatically scored as incorrect.

Here are some items you will need to bring with you to the exam site:

- A valid government or school-issued photo ID
- Your test center admissions ticket
- A watch (not all test sites have clocks)

> *Note:* Books, calculators, and other reference materials are not allowed in the exam room. The exceptions to this list are English-Foreign dictionaries or glossaries. All items will be inspected before and after testing.

WHAT TO EXPECT DURING THE EXAM

When taking a CBT exam, as soon as you are seated in the testing center, you will be given a brief tutorial to acquaint you with the computer-delivered test, prior to taking your certification exam(s). The CBT exams allow you to select only one answer per question. You can also change your answers as many times as you like. When you select a second answer choice, the CBT will automatically unselect your first answer choice. If you want to skip a question to return to later, you can utilize the "flag" feature, which will allow you to quickly identify and review questions whenever you are ready. Prior to completing your exam, you will also be provided with an opportunity to review your answers and address any unanswered questions.

TESTING TIME

Each individual ASE CBT exam has a fixed time limit. Individual exam times will vary based upon exam area and will range anywhere from a half hour to two hours. You will also be given an additional 30 minutes beyond what is allotted to complete your exams to ensure you have adequate time to perform all necessary check-in procedures, complete a brief CBT tutorial, and potentially complete a post-test survey.

You can register for and take multiple CBT exams during one testing appointment. The maximum time allotment for a CBT appointment is four and a half hours. If you happen to register for so many exams that you will require more time than this, your exams will be scheduled into multiple appointments. This could mean that you have testing on both the morning and afternoon of the same day, or they could be scheduled on different days, depending on your personal preference and the test center's schedule.

It is important to understand that if you arrive late for your CBT test appointment, you will not be able to make up any missed time. You will only have the scheduled amount of time remaining in your appointment to complete your exam(s).

Also, while most people finish their CBT exams within the time allowed, others might feel rushed or not be able to finish the test, due to the implied stress of a specific, individual time limit allotment. Before you register for the CBT exams, you should review the number of exam questions that will be asked along with the amount of time allotted for that exam to determine whether you feel comfortable with the designated time limitation or not.

As an overall time management recommendation, you should monitor your progress and set a time limit you will follow with regard to how much time you will spend on each individual exam question. This should be based on the total number of questions you will be answering.

Also, it is very important to note that if for any reason you wish to leave the testing room during an exam, you must first ask permission. If you happen to finish your exam(s) early and wish to leave the testing site before your designated session appointment is completed, you are permitted to do so only during specified dismissal periods.

UNDERSTANDING HOW YOUR EXAM IS SCORED

You can gain a better perspective about the ASE certification exams if you understand how they are scored. ASE exams are scored by an independent organization having no vested interest in ASE or in the automotive industry. With CBT exams, you will receive your exam scores immediately.

Each question carries the same weight as any other question. For example, if there are 50 questions, each is worth 2 percent of the total score. The passing grade is 70 percent. That means you must correctly answer 35 out of the 50 questions to pass the exam.

Your exam results can tell you

- Where your knowledge equals or exceeds that needed for competent performance, or
- Where you might need more preparation.

Your ASE exam score report is divided into content "task" areas; it will show the number of questions in each content area and how many of your answers were correct. These numbers provide information about your performance in each area of the exam. However, because there may be a different number of questions in each content area of the exam, a high percentage of correct answers in an area with few questions may not offset a low percentage in an area with many questions.

It should be noted that one does not "fail" an ASE exam. The technician who does not pass is simply told "More Preparation Needed." Though large differences in percentages may indicate problem areas, it is important to consider how many questions were asked in each area. Since each exam evaluates all phases of the work involved in a service specialty, you should be prepared in each area. A low score in one area could keep you from passing an entire exam. If you do not pass the exam, you may take it again at any time it is scheduled to be administered.

There is no such thing as average. You cannot determine your overall exam score by adding the percentages given for each task area and dividing by the number of areas. It does not work that way because there generally is not the same number of questions in each task area. A task area with 20 questions, for example, counts more toward your total score than a task area with 10 questions.

Your exam report should give you a good picture of your results and a better understanding of your strengths and areas needing improvement for each task area.

Types of Questions on an ASE Exam

Understanding not only what content areas will be assessed during your exam, but how you can expect exam questions to be presented will enable you to gain the confidence you need to successfully pass an ASE certification exam. The following examples will help you recognize the types of question styles used in ASE exams and assist you in avoiding common errors when answering them.

Most initial certification tests are made up of between 40 and 80 multiple-choice questions. The five-year recertification exams will cover the same content as the initial exam; however, the actual number of questions for each content area will be reduced by approximately one-half. Refer to Section 4 of this book for specific details regarding the number of questions to expect to receive during the initial Electrical/Electronic Systems (T6) certification exam.

Multiple-choice questions are an efficient way to test knowledge. To correctly answer them, you must consider each answer choice as a possibility, and then choose the answer choice that *best* addresses the question. To do this, read each word of the question carefully. Do not assume you know what the question is asking until you have finished reading the entire question.

About 10 percent of the questions on an actual ASE exam will reference an illustration. These drawings contain the information needed to correctly answer the question. The illustration should be studied carefully before attempting to answer the question. When the illustration is showing a system in detail, look over the system and try to figure out how the system works before you look at the question and the possible answers. This approach will ensure you do not answer the question based upon false assumptions or partial data, but instead have reviewed the entire scenario being presented.

MULTIPLE-CHOICE/DIRECT QUESTIONS

The most common type of question used on an ASE exam is the direct multiple-choice style question. This type of question contains an introductory statement, called a stem, followed by four options: three incorrect answers, called distracters, and one correct answer, the key. When the questions are written, the point is to make the distracters plausible to draw an inexperienced technician to inadvertently select one of them. This type of question gives a clear indication of the technician's knowledge.

Here is an example of a direct style question:

1. The low-voltage disconnect (LVD) system opens (turns off power) when the battery voltage drops to what level?

    A. 8.4 volts

    B. 12.6 volts

    C. 10.4 volts

    D. 9.6 volts

Answer A is incorrect. A level of 8.4 volts would be too low to allow the engine to start.

Answer B is incorrect. A level of 12.6 volts is the fully charged voltage on truck batteries.

Answer C is correct. Most LVD systems will open when battery voltage drops to 10.4 volts. This is the factory preset level and can be adjusted to suit individual needs.

Answer D is incorrect. A level of 9.6 volts would be too low to allow the engine to start. The specification for performing a battery load test is 9.6 volts.

TECHNICIAN A, TECHNICIAN B QUESTIONS

This type of question is usually associated with an ASE exam. It is, in fact, two true-false statements grouped together, such as: "Technician A says…" and "Technician B says…", followed by "Who is correct?"

In this type of question, you must determine whether either, both, or neither of the statements are correct. To answer this type of question correctly, you must carefully read each technician's statement and judge it on its own merit.

Sometimes this type of question begins with a statement about some analysis or repair procedure. This statement provides the setup or background information required to understand the conditions about which Technician A and Technician B are talking, followed by two statements about the cause of the concern, proper inspection, identification, or repair choices.

Analyzing this type of question is a little easier than the other types because there are only two ideas to consider, although there are still four choices for an answer.

Technician A, Technician B questions are really double true-or-false questions. The best way to analyze this type of question is to consider each technician's statement separately. Ask yourself, "Is A true or false? Is B true or false?" Once you have completed an individual evaluation of each statement, you will have successfully determined the correct answer choice for the question, "Who is correct?"

An important point to remember is that an ASE Technician A, Technician B question will never have Technicians A and B directly disagreeing with each other. That is why you must evaluate each statement independently.

An example of a Technician A/Technician B style question looks like this:

1. A truck electrical system is being repaired. Technician A says that wiring schematics give exact details of the location of all electrical components on a truck. Technician B says that a wiring schematic will usually contain pin-out test procedures that can be used to troubleshoot many electrical faults. Who is correct?
    A. A only
    B. B only
    C. Both A and B
    D. Neither A nor B

Answer A is incorrect. A wiring schematic is a drawing that shows the wires and components in a circuit using symbols to represent the components.

Answer B is incorrect. A wiring schematic does not typically contain a pin-out procedure.

Answer C is incorrect. Neither Technician is correct.

Answer D is correct. Neither Technician is correct. A wiring schematic is a drawing that shows the wires and components in a circuit using symbols to represent the components. These schematics are very useful in the diagnosis of electrical problems on trucks because a technician can use them to decide where to begin his/her testing steps.

EXCEPT QUESTIONS

Another type of question used on ASE exams contains answer choices that are all correct except for one. To help easily identify this type of question, whenever it is presented in an exam, the word "EXCEPT" will always be displayed in capital letters. Furthermore, a cautionary statement will alert you to the fact that the next question is different from the ones otherwise found in the exam. With the EXCEPT type of question, only one incorrect choice will actually be listed among the options, and that incorrect choice will be the key to the question. That is, the incorrect statement is counted as the correct answer for that question.

Be careful to read these question types slowly and thoroughly; otherwise you may overlook what the question is actually asking and answer the question by selecting the first correct statement.

An example of this type of question would appear as follows:

1. All of the following conditions can cause reduced current flow in an electrical circuit EXCEPT:

   A. Loose terminal connections
   B. A hot-side wire that is rubbing a metal component
   C. Corrosion inside the wire insulation
   D. Wire that is too small

Answer A is incorrect. Loose terminal connections will reduce the current flow because of the added electrical resistance.

Answer B is correct. A hot-side wire that rubs a metal component will cause a large increase in current flow and would likely open a circuit protection device.

Answer C is incorrect. Wire corrosion will reduce the current flow because of the added electrical resistance.

Answer D is incorrect. A wire that is too small will reduce the circuit's ability to carry the correct current because of the limited surface area to allow the electrons to flow.

LEAST LIKELY QUESTIONS

LEAST LIKELY questions are similar to EXCEPT questions. Look for the answer choice that would be the LEAST LIKELY cause of the described situation. To help easily identify this type of question, whenever they are presented in an exam, the words "LEAST LIKELY" will always be displayed in capital letters. In addition, you will be alerted before a LEAST LIKELY question is posed. Read the entire question carefully before choosing your answer.

An example of this type of question is shown below:

1. Which of the following devices would be LEAST LIKELY used as a circuit protection device?

   A. Maxi-fuse
   B. Circuit breaker
   C. Relay
   D. Positive temperature coefficient (PTC) thermistor

Answer A is incorrect. Maxi-fuses have a metallic strip that burns up when the current flow in the circuit rises above the maxi-fuse rating. Once the fuse opens, it must be replaced to have continued electrical operation.

Answer B is incorrect. Circuit breakers are made of a bi-metallic strip that bends when it gets hot. High current flow in a circuit will cause the circuit breaker to heat up, which causes the device to open its contacts.

Answer C is correct. A relay is an electromagnetic switch that allows a large current circuit to be controlled by a small current circuit.

Answer D is incorrect. Some late-model trucks use positive temperature coefficient (PTC) thermistors as a circuit protection device. PTCs are electronic devices that have low resistance when they are at ambient temperature. The resistance of these devices increases as their temperature rises. This rising resistance limits current flow in the circuit they are protecting.

SUMMARY

The question styles outlined above are the only ones you will encounter on any ASE certification exam. ASE does not use any other types of question styles, such as fill-in-the-blank, true/false, word-matching, or essay. ASE also will not require you to draw diagrams or sketches to support any of your answer selections, although any of the above described question styles may include illustrations, charts, or schematics to clarify a question. If a formula or chart is required to answer a question, it will be provided for you. You may in rare cases be required to solve a simple math problem, so bringing a simple pocket calculator to the test session might be a good idea.

Task List Overview

INTRODUCTION

This section of the book outlines the content areas or *task list* for this specific certification exam, along with a written overview of the content covered in the exam.

The task list describes the actual knowledge and skills necessary for a technician to successfully perform the work associated with each skill area. This task list is the fundamental guideline you should use to understand what areas you can to expect to be tested on, as well as how each individual area is weighted to include the approximate number of questions you can expect to be given for that area during the ASE certification exam. It is important to note that the number of exam questions for a particular area is to be used as a guideline only. ASE advises that the questions on the exam may not equal the number of specifically listed on the task list. The task lists are specifically designed to tell you what ASE expects you to know how to do and to help prepare you to be tested.

Similar to the role this task list will play in regard to the actual ASE exam, Delmar, Cengage Learning has developed six preparation exams, located in Section 5 of this book, using this task list as a guide. It is important to note that although both ASE and Delmar, Cengage Learning use the same task list as a guideline for creating these test questions, none of the test questions you will see in this book will be found in the actual, live ASE exams. This is true for any test preparatory material you use. Real exam questions are *only* visible during the actual ASE exams.

Task List at a Glance

The Electrical/Electronic Systems (T6) task list focuses on five core areas, and you can expect to be asked a total of approximately 50 questions on your certification exam, broken out as outlined:

- A. General Electrical/Electronic System Diagnosis (14 questions)
- B. Battery and Starting System Diagnosis and Repair (11 questions)
- C. Charging System Diagnosis and Repair (7 questions)
- D. Lighting Systems Diagnosis and Repair (6 questions)
- E. Related Vehicle Systems Diagnosis and Repair (12 questions)

Based upon this information, the following graphic is a general guideline demonstrating which areas will have the most focus on the actual certification exam. This data may help you prioritize your time when preparing for the exam.

> *Note:* There could be additional questions that are included for statistical research purposes only. Your answers to these questions will not affect your test score, but since you do not know which ones they are, you should answer all questions in the test. The five-year Recertification Test will cover the same content areas as those listed above. However, the number of questions in each content area of the Recertification Test will be reduced by one-half.

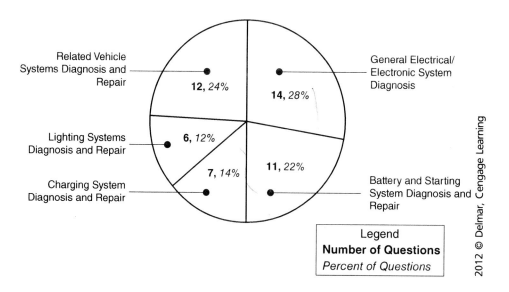

ELECTRICAL/ELECTRONIC SYSTEMS (TEST T6) TASK LIST

A. General Electrical/Electronic System Diagnosis (14 questions)

1. Check applied voltages, circuit voltages, and voltage drops in electrical/electronic circuits using digital multimeter (DMM), or appropriate test equipment.

Voltage tests are measured with the test leads of the DMM in parallel to the component or circuit being tested. In testing applied voltage, the negative or black test lead of the DMM is connected to a battery or chassis ground. The positive red lead is then probed near the power source (e.g., a switch or fuse) to determine if the circuit is receiving the proper voltage. A handy feature of a DMM is that polarity is not important. If the leads are reversed, the display will simply include a minus sign in front of the reading. Circuit voltages are tested much the same way. Take the positive test probe and go from one circuit's power source to the next. Generally, the applied voltages should all be the same.

When selecting a DMM to use for testing electronic circuits, it is recommended that it have a 10 megohm or higher impedance. This is necessary to limit the effect the unit might have on accuracy when testing low current flow circuits. For this reason, never use a test light to check for voltage in an electronic circuit. The lamp in the test light will draw too much current and this can damage the circuit integrity.

If it is necessary to test for voltage in an energized circuit that is in operation, a convenient way to do this is with a back-probe tool, also known as a "spoon," at various connectors in a particular circuit. This DMM accessory makes it possible to probe into the connector from behind without disconnecting it.

2. Check current flow in electrical/electronic circuits and components using a digital multimeter (DMM), clamp-on ammeter, or appropriate test equipment.

Current flow tests are used when a circuit is suspected of having higher than normal current flow, such as a dragging blower motor, or a circuit with a low-resistance short to ground.

A technician that has a good knowledge of what normal amperage should be can use this knowledge when troubleshooting electrical circuits.

In testing current flow with an ammeter or DMM, it is important to remember that the test leads are connected in series with the circuit being tested, usually at a point near the power source. The circuit must be interrupted at some point to allow the connection of the test leads.

Most DMMs have a 10–20 amp limit when measuring amperage directly through the meter. Any greater current flow will blow its fuse. If you suspect that the circuit carries more than that, then a safe way to test for current flow would be to use a current clamp. This device simply clamps over the wire being tested and determines current flow by measuring the strength of the magnetic field surrounding the wire. While it is extremely handy to use, it is not as accurate as routing all the current through the DMM, especially in circuits flowing less than 10 amps.

> *Remember:* Low current flow usually is a result of excessive resistance in a circuit or low voltage. Higher than normal current flow can generally be traced to excessive applied voltage or a shorted component or wire.

3. Check continuity and resistance in electrical/electronic circuits and components using a digital multimeter (DMM), or appropriate test equipment.

Checking continuity in an electrical circuit is one of the most common troubleshooting tests a technician performs. Auto-ranging DMMs do not have to be adjusted to the range that is being measured While the purpose of this test is to make sure that a complete current path exists in the circuit being tested, it is important to remember that it is not an accurate indication of circuit performance (e.g., excessive resistance). A continuity test is most useful to quickly differentiate one circuit from another, such as trying to locate a specific contact in a multiple pin connector.

Continuity tests can be made in a number of different ways. If the circuit is energized, a 12/24 volt test light or voltmeter can be used to check for voltage at various test points. Both tools accomplish this by energizing the circuit with low potential current to determine that the circuit can be closed. Many DMMs have a separate feature on them that will allow continuity tests to be made simply by listening for an audible beep. This is handy because multiple tests can be made rather quickly, without having to look constantly at the display for a resistance value.

When testing circuits that include electronic control modules (ECMs), it is essential to observe the original equipment manufacturer's (OEM's) diagnostic procedure to avoid possible damage to the processor.

Resistance checks are typically made when a circuit has unwanted voltage drops or low current flow. An ohmmeter is a device that circulates a small current through a circuit when it is not energized and then measures the voltage drop through it. It displays this resistance (or restriction to current flow) in units known as ohms. The lower the resistance value, the less restriction to electron flow there will be in a circuit. If the meter reads infinity (or a high flashing number on most DMMs), this means that the circuit is open. Except for where resistance is built in to a circuit, such as a blower motor resistor or a load itself, generally speaking the lower resistance a circuit has the better. For example, when testing a length of wire or a fuse, most will test very near zero ohms. When testing with a DMM, always be sure to zero the meter first to compensate for any resistance present in the test leads, especially when testing low-resistance components or circuits.

When making tests with a DMM that is not auto-ranging, be sure to select a range that will provide accuracy. If you are using an analog meter, set the meter to a range setting that will put the needle in roughly the middle of the scale for the component you are testing.

Resistance checking of specific components is generally used when a manufacturer specifies a certain test value, such as a fuel level sending unit. Some components, such as light bulbs and glow plugs, do not lend themselves to resistance testing because their resistance changes as they heat up. Also, large-diameter conductors such as battery cables cannot be reliably tested with an ohmmeter because it cannot circulate enough current to simulate actual operating conditions and identify resistance.

4. Find shorts, grounds, and opens in electrical/electronic circuits.

A short circuit is defined as one where the current flow is allowed to ground at a point other than where it was designed, such as a bare wire rubbing against the frame. An open circuit is usually caused by a broken wire or other component not making the necessary connection to complete a circuit, stopping current flow in that portion of the circuit.

Finding shorts is best done with an ohmmeter, because in a live circuit a fuse or breaker will continually blow. Do not install a larger fuse; it will risk melting a bundle of wires. Divide the circuit into small sections at various connectors (where applicable) while testing for continuity to ground (there should be none) or visually inspect the harness for rub or pinch points.

Locating opens can sometimes be more difficult because sometimes the damage is not visually apparent. A good way to test for an open would be to apply voltage at one end and then probe at each succeeding connection downstream until you find no reading. Alternatively, an ohmmeter can be used to do the same thing when the circuit is not energized.

It is important to remember when testing at connectors, especially smaller ones designed for electronic circuits, that you do not damage the contacts when probing into them, especially female ones. Spreading the tongs on a female contact while testing can create problems. Always use an appropriate adapter for testing these types of contacts.

5. Diagnose key-off battery drain (parasitic draw) problems; determine needed repairs.

While dead batteries are not always caused by a key-off battery drain problem, it is a good idea before you start troubleshooting to know the possible causes for this problem and how the system is constructed. Before the widespread use of electronics in vehicles, many systems had zero current draw with the key off. Today, with so many vehicles having multiple control modules, testing for battery drain will likely involve disconnecting ECMs or isolating circuits.

In older vehicles, disconnect the negative battery cable and connect an ammeter in series with it and the ground post. Make sure the key is off and all loads, such as dome lights, are turned off. There should be no current draw. If there is still a current draw, an easy way to isolate the problem is to start pulling fuses one at a time until the draw stops. Another possible draw is through a defective diode in the alternator. Disconnect the positive lead at the alternator to locate this potential problem.

On vehicles with ECMs, use a milliamp scale on a DMM and consult the manufacturer's specifications to ensure that a draw is within parameters: Most will draw well under

50 milliamps with the key off. If it measures higher than specified, it might be necessary to disconnect the ECM's power supply to be sure that there is not another component in the vehicle causing the additional draw. Be sure to allow time for the ECM to power down and enter "sleep mode" before taking a final reading.

6. Inspect and test fusible links, circuit breakers, fuses, and other circuit protection devices, including reset when required.

A fuse is an electrical safety device. When it blows, it is because of a current overload somewhere in the circuit. Always repair the problem: Never install a fuse of a higher rating. Also, learn to identify the reason for a fuse failure. If the metal filament in the center of the fuse melts, it is caused by a current overload. On older glass-style fuses, melted end caps are caused by poor or corroded contacts in the fuse holder itself, not a current overload.

A circuit breaker performs the same function as a fuse; however, it has a feature that allows it to be reset after tripping, usually automatically. Most circuit breakers can be identified as a small rectangular box with two studs attached to it. Their maximum current rating is stamped on the housing. Two types of circuit breaker are used in truck electrical circuits. SAE #1 circuit breakers cycle when overloaded. SAE #2 circuit breakers trip when overloaded and do not reset until the circuit is opened or the reset button is depressed. Circuit breakers are handy test devices to have in a tool box. When testing a circuit that continually blows conventional fuses, installing a circuit breaker into the circuit temporarily with jumper wires saves both time and money.

Fusible links are short sections of wire designed to melt and open a circuit in case of overload. They are usually installed near a power source (e.g., battery or starter solenoid) and are normally two to four wire gauge sizes smaller in diameter than the circuit they are protecting. When they do melt, the insulation usually bubbles (but not always), making them difficult to troubleshoot. The fuse link has a special high-temperature insulation designed not to separate during an overload. Sometimes the easiest way to test these devices is to simply give them a good tug at either end. If it stretches, it is defective.

Many manufacturers are now using what are known as maxi-fuses in place of fusible links, which are also usually located near the battery or main power distribution bus. They simply look like conventional fuses, although much larger than standard. They are also much easier to install and troubleshoot.

To test any of the previously mentioned circuit safety devices, remove the component and test for resistance using an ohmmeter. A good component should show very low resistance.

7. Inspect and test spike suppression diodes/resistors and capacitors.

A diode is simply an electrical "check valve" that allows current to flow in only one direction. A symbol of a diode looks like an arrow with a line drawn perpendicular to its point. A typical use of a diode is in an alternator, where it performs the task of converting AC voltage into DC by simply preventing output to the battery during the negative cycle of the sine wave. Diodes can also be used to prevent backfeeding of current from the alternator excitation circuit to key switch circuits once the engine starts.

Testing diodes is simple. Use an ohmmeter and check for continuity through both directions. If it is good, current will flow in one direction, but not the other.

Spike suppression devices can be in the form of diodes, resistors, or capacitors. Their purpose is to absorb or redirect a voltage spike that might come from a collapsing magnetic field, such as when the air conditioner (A/C) compressor clutch coil is switched off. By installing such a device in parallel with the coil, the voltage spike is directed back to the clutch coil and prevented from damaging sensitive components, such as ECMs.

8. Inspect and test relays and solenoids (including solid-state devices).

A relay is defined as a switching device that uses a small amount of current to control a larger one. A solenoid is a device that performs mechanical movement when electrically energized, such as a fuel shutoff solenoid. A solenoid can also incorporate a relay function. A good example of this would be a starter solenoid, which not only moves the starter drive pinion into mesh with the flywheel ring gear, but also makes the high-current connection between the battery and the starter field coils.

Most mini-relays have four or five terminals. The two small terminals (often specified as #85 and #86) are used to energize the coil that creates the magnetic attraction necessary to cause a connection between the high amperage switch contacts. Two other terminals make the high-amperage power in (terminal #30) and power out (terminal #87) connection. Sometimes a fifth terminal (marked #87a) is used as a normally closed (NC) contact, as opposed to #87 being normally open (NO), which closes when the relay energizes. Note that the physical size of the power terminals (#30, #87, and #87a) may or may not be larger than the control terminals (#85 and #86), depending on the amperage capacity of the relay.

In operation, when a relay is signaled to close the high-current contacts, a small amount of current is fed through terminals #86 and #85, one being battery positive and the other battery ground. This signal can come from an ECM, key switch, or other low-current switching device. The positive side of the high-current contact (#30) is then connected to the load side (#87), completing the circuit to the high-current device, such as a horn or multiple light circuits. Some relays do not have the same code numbers, but function using the same principles.

Some manufacturers use solid-state relays in different areas of the truck. The operation of these devices is much like an electronic fuel injector. These components can be totally off or totally on, like regular relays. What differentiates them is their ability to be turned on to a certain percentage with a pulse-width modulated signal. The advantage of these devices is that they can control the current flow to the output device, resulting in a variable speed device. A scan tool is necessary to troubleshoot these solid-state relays. The scan tool can retrieve trouble codes as well as perform output tests on these circuits during the troubleshooting process.

9. Read and interpret electrical schematic diagrams and symbols.

Technicians should be familiar with most of the standard symbols for electrical circuit schematics. Valley Forge schematic symbols were once considered standard in the United States and Canada, but due to the widespread ownership of our trucking OEMs by European companies, some manufacturers now use European schematic symbols and architecture. The figure below shows some of the symbols used by International Trucks. Other manufacturers will use similar, although not necessarily identical, graphics. Pay special attention to the symbols used for grounds, connections, fuses, diodes, switches, relays, and twisted pairs (data bus).

See the chart below for some of the common symbols used in wiring diagrams.

SYMBOLS USED IN WIRING DIAGRAMS			
+	Positive	(temperature switch symbol)	Temperature switch
—	Negative	(diode symbol)	Diode
(ground symbol)	Ground	(zener diode symbol)	Zener diode
(fuse symbol)	Fuse	(motor symbol)	Motor
(circuit breaker symbol)	Circuit breaker	→〉—C101	Connector 101
(condenser symbol)	Condenser	→	Male connector
(ohms symbol)	Ohms	〉—	Female connector
(fixed value resistor symbol)	Fixed value resistor	—●	Splice
(variable resistor symbol)	Variable resistor	S101	Splice number
(series resistors symbol)	Series resistors	(thermal element symbol)	Thermal element
(coil symbol)	Coil	(multiple connectors symbol)	Multiple connectors
(open contacts symbol)	Open contacts	88:88	Digital readout
(closed contacts symbol)	Closed contacts	(single filament bulb symbol)	Single filament bulb
(closed switch symbol)	Closed switch	(dual filament bulb symbol)	Dual filament bulb
(open switch symbol)	Open switch	(light-emitting diode symbol)	Light-emitting diode
(ganged switch symbol)	Ganged switch (N.O.)	(thermistor symbol)	Thermistor
(single pole double throw switch symbol)	Single pole double throw switch	(PNP transistor symbol)	PNP bipolar transistor
(momentary contact switch symbol)	Momentary contact switch	(NPN transistor symbol)	NPN bipolar transistor
(pressure switch symbol)	Pressure switch	(gauge symbol)	Gauge

10. Diagnose failures in the data communications bus network; determine needed repairs.

All trucks that have one or more on-board computers will also have a network that allows the computers to communicate with each other. This network also allows for scan tool communications. Typically, this network consists of two wires that are connected to the various modules and the data connector in parallel. The process of pulsing voltage to send signals is called multiplexing or bussing. The bus network wires are continuously twisted in order to help resist radio frequency interference (RFI) and electromagnetic interference (EMI) from entering this system. This method of sharing data among modules helps eliminate redundant wiring to the modules that need the same sensor information.

The technician can communicate with the truck's various control modules by connecting a scan tool to the truck. A 6-pin or 9-pin data link connector is the location where the technician connects the scan tool. Late-model trucks will typically have a data network called the J1939 network. This network is very effective because it can be used with up to 30 computers and communicates at very high speeds. The J1939 network uses two terminating resistors that connect in parallel with the two-wire system. It is sometimes necessary for the technician to measure the resistance of the data network at the 9-pin connector. The negative battery cable should be disconnected prior to measuring this resistance. The correct resistance of the data network should be approximately 60 ohms. If this measurement is incorrect, the technician should inspect for problems concerning the terminating resistors.

Problems that can occur in the bus communication are opens, shorts, and unwanted resistance. Diagnosing these problems involves using the DMM in conjunction with a diagnostic trouble chart. The scan tool becomes less valuable when a communication problem occurs, because it will not be able to receive data in the event of a wire failure.

11. Diagnose vehicle electronic control systems using appropriate diagnostic tools, software, and service information; check and record diagnostic codes; determine needed repairs.

Late-model trucks are equipped with several on-board computers that control many of the systems on the truck. Examples of systems that are controlled by computers are the engine, transmission, antilock brakes, collision avoidance system, and the instrument panel. It is important to remember that all computer systems have some key characteristics in common with one another. Computers need to be connected to input devices such as switches and sensors. Computers need to have processing capability in order to analyze the input data to determine how to react to the information. Computers also need to be connected to output components in order to perform work.

In order to repair systems that use computers to control their operations, a technician needs to be able to communicate with the computer in question. One way to achieve communication with the computer on the truck is to use a scan tool. A scan tool is a portable computer that is used to retrieve codes, view live data, and run output tests on the truck's computer systems. Scan tools can be stand-alone units or installed on a laptop computer. In either case, the scan tool connects to the truck using a cord with a data connector that connects to the data connector on the truck. See Task A.12 for more information.

Many on-board computer systems also have the ability to display trouble codes using flash codes. This method typically requires the technician to press a switch or ground a wire in order to put the system in diagnostic mode. The code is displayed by causing a light or indicator to flash a numbered sequence. The technician counts the number of times the light flashes and interprets the code.

The technician uses the information retrieved from the truck's computer system to find out where to begin the diagnostic process. A database must be consulted to find the correct troubleshooting information. Most databases used today are computer-based or web-based. The database assists the technician by providing flowcharts, diagnostic routines, and technical service bulletins (TSBs).

12. Connect diagnostic tool to vehicle; access and verify parameters and calibration settings; perform updates as needed.

The scan tool is typically connected to the truck using a long wire with the correct diagnostic connector. The most common diagnostic connector on late-model trucks is a round (nine-pin) connector. This connector is usually located somewhere inside the cab of the truck.

The scan tool has extensive diagnostic power for troubleshooting and repairing trucks. It is used to retrieve and clear diagnostic trouble codes (DTCs), view live data, and perform output tests. It is always advisable to connect the scan tool early in the diagnostic routine if one is available. The technician can investigate if DTCs are present, as well as view live data from the many switches and sensors on the truck. If DTCs are present, the technician should use the manufacturer's database to find the appropriate diagnostic routine for diagnosing the code. See Task A.11 for more information.

Many scan tools also can be used to view custom parameters and change calibration settings. There are several parameters on a truck that can be programmed to suit the owner. Examples of the settings that can be viewed and calibrated include maximum vehicle speed, maximum cruise control speed, and maximum idle time. The technician can also view parameters that show how the truck has been driven. Examples of this data include the average fuel economy, the number of hours the truck has idled, and the percentage of time the throttle has been at full range.

Many of the computers on the truck can be reprogrammed with the scan tool. This function is performed at times when the manufacturer releases a new programming level that improves the performance of a system or function. This information is available to the technician in the electronic database that is used to find repair information.

B. Battery and Starting System Diagnosis and Repair (11 questions)

1. Determine battery state of charge by measuring terminal post voltage using a digital multimeter (DMM).

A rough estimate of battery state of charge can be determined by measuring its open circuit voltage. With the battery under no load, measure voltage across the terminals. If it is 12.6 volts or more, it is considered fully charged. If the battery has been recently charged (whether inside or outside the vehicle), draw off the surface charge using this same method. Allow the battery to sit for 15 minutes before testing, then test the open circuit voltage. Any reading less than 12.6 volts indicates that the battery should be charged.

Battery state of charge can also be determined by using a hydrometer. A specific gravity reading of 1.265 or higher at 80° F (26.6° C) indicates a fully charged battery. A severely discharged battery will show a reading of around 1.120 or so. A reading of 1.200 would indicate a battery that is 50 percent charged. Temperature corrections must be made to the readings for batteries not within 10° of 80° F (or between 21.1° C and 32.2° C).

Battery state of charge can also be determined by an open circuit voltage (OCV) test across the battery terminals. When the OCV has stabilized, the following readings can be used:

- 12.6 volts or more   Fully charged
- 12.4 volts   75% charged
- 12.2 volts   50% charged
- 12.0 volts   25% charged
- 11.7 volts or less   Fully discharged

2. Perform battery tests (load and capacitance); determine needed service.

To properly perform a battery load test, first determine whether the battery is fully charged. There is no point in testing a partially charged battery when that condition is caused by charging system problems, because it will fail. First determine the battery state of charge according to the steps outlined in Task B.1.

When you are sure the battery is ready to be load tested, draw off the surface charge if the battery has just been charged by either an alternator or battery charger. Load the battery by either cranking the starter for 15 seconds or drawing 300 amps with the load tester for 15 seconds. Allow the battery to sit for a few minutes.

To perform the load test, first determine the rating of the battery. This is usually expressed in cold cranking amps (CCA), although some older batteries might be rated in ampere-hours (AH). Draw the battery down with the load tester at a rate equal to one-half the CCA or three times the AH. Hold this load for 15 seconds. At the end of 15 seconds, with the load still applied, note battery voltage. A reading over 9.6 volts at 70° F (21.1° C) means the battery has passed the test. A battery that passes at 10.7 volts versus one that passes at 9.7 volts is the better of the two. One that barely passes will not last long. Keep this in mind in severe weather climates.

Some truck OEMs recommend capacitance testing of batteries; a number of different OEM instruments are available that will perform this test. The test instrument outputs a low potential AC signal and measures the return pulse. The instrument display is idiot-proof and usually reads in terms of OK or not OK. The danger of using this method is that some good batteries could possibly be rejected as not OK.

3. Inspect, clean, service, or replace battery, cables, and terminal connections.

Start any battery service routine by wearing eye protection. Inspect the top of the battery case for a buildup of dirt and moisture that can cause a low-amperage current draw across the top of the battery. This can be checked by taking one probe of a DMM and dragging it across the top of the battery while holding the other probe on one of the posts. Any significant reading indicates a low-current short between the two battery terminals, which can result in a low or dead battery over a period of time. A battery is best cleaned with a water and baking soda solution.

Battery cables and their terminal ends are a frequent source of problems. Many times they are the cause of a no-start or a sluggish starting complaint. A simple voltage drop test (see Task B.9) will quickly identify the connection(s) with excessive resistance. Tapered battery posts and cable ends are best cleaned with a type of scraper tool that actually peels away all the old corrosion down to bright shiny metal. Flat, screw-type posts are more difficult to clean; however, it is just as important that they be clean in order to transfer current with minimal voltage drop.

After reinstalling battery cable ends, coat the terminals with grease or petroleum jelly, or use a spray marketed for this purpose, to resist corrosion. Protective pads that go under the tapered terminals serve the same function.

When removing a battery and/or cables, always remove the negative cable first, and reconnect it last. This will help to prevent arcing and a possible battery explosion should your wrench come into contact with ground when loosening the positive cable.

If a conventional battery is low on electrolyte level, add distilled water only. If a battery needs to have distilled water added on a regular basis, it would be wise to check the charging system for a potential problem.

4. Inspect, clean, repair or replace battery boxes, mounts, and hold downs.

The life of a battery depends in large part on the way it is secured in its mounts. A battery with missing straps or mounts will bounce around and eventually damage the internal separator plates, which may cause an internal short. Similarly, overtightened hold down straps may cause the case to crack.

Inspect the battery box when the battery is removed. Clean away any dirt, rust, and corrosion to help combat surface discharge when the battery is reinstalled.

5. Charge battery using appropriate method for battery type.

The amount of time it takes to properly charge a battery depends on its state of charge. The other factor to consider is whether to charge it at a fast or slow rate. In most cases, a slow charge rate allows for maximum battery life and performance.

When slow charging, a 5–10 amp rate is usually sufficient, although a full charge may take overnight if the battery charge is low. If a fast charge is required, certain precautions must be taken. Never allow more than a 50–60 amp charge rate, and then monitor the electrolyte temperature to make sure it does not exceed 125° F (51.67° C). If the specific gravity reaches 1.225 during the fast charge, reduce the charge rate accordingly. If gassing or spewing of electrolyte occurs, reduce the charge rate. Never charge a frozen battery until it is brought up to room temperature.

Batteries should only be charged in well-ventilated areas. In addition, batteries should not be charged near heat sources or sources of ignition. If the caps are removed during charging, cover the top of the battery with a moist rag. You can also monitor the specific gravity of the battery during charging to determine its state of charge. A reading of 1.265 indicates a full charge. Alternately, an ammeter on the charger that slowly drops to zero or near zero indicates a fully charged battery. A battery that will not accept a charge from the start (zero amps) is likely to be highly sulfated and should be disposed of according to Universal Waste guidelines.

6. Jump start a vehicle using jumper cables and a booster battery or appropriate auxiliary power supply.

Jump starting a vehicle with a dead battery is a simple procedure. It can be dangerous, however. Proper procedures must be followed to ensure that a spark is not generated that can cause an explosion. A battery that is low due to prolonged cranking is likely generating explosive vapors near the vent caps and a spark at this location could be dangerous.

Always wear eye protection when jump starting a vehicle. Then, with both vehicle engines off, connect the positive terminals of both batteries with one cable. Connect the ground booster cable to the negative battery of the booster vehicle and make the last connection to the negative battery terminal(s) of the dead vehicle. This will prevent sparks in the area of the dead battery.

Start the engine of the booster vehicle. If the jumper cables are of a generous size, crank and start the dead engine immediately. If the cables are small, or you are using a fast/ boost charger, allow the connection to remain for a few minutes while the dead battery recharges. Then, with the help of the booster battery, crank and start the engine. Remove the cables in the reverse order of installation.

7. Diagnose low voltage disconnect (LVD) systems; determine needed repairs.

Low voltage disconnect (LVD) systems are installed on many trucks to help prevent the main battery set from discharging too much in starting the truck. The module for the LVD system is installed in series between the battery set and the truck power distribution center. This module monitors battery voltage and opens the circuit when the voltage falls below a preset limit. The circuit remains open until battery voltage rises to a higher preset limit. The disconnect voltage is typically around 10.4 volts and the reconnect voltage is around 12.2 volts. These parameters can be changed, depending on the needs of the truck operator.

8. Test/monitor battery and starting system voltage during cranking; determine needed repairs.

Battery voltage during engine cranking can be monitored using a DMM. Since all late-model trucks have computers, it is crucial that cranking voltage be sufficient to allow them to work correctly. As in the past, low cranking voltage will damage the starter due to increased amperage. The battery voltage should not drop below approximately 10.5 volts while cranking.

9. Perform starting circuit voltage drop tests; determine needed repairs.

Problems related to the cranking system are common vehicle malfunctions. It is important to know how to perform a few simple tests in this area to solve the problem quickly and effectively. A cranking system voltage drop test identifies high resistance in the cranking circuit that can cause a slow or no-start condition. Every connection and conductor in the circuit, from the battery to the solenoid through the ground path, is a potential problem area.

To test for voltage drop in the positive battery cable, set a DMM to the V-DC scale and attach one probe to the positive terminal post of the battery (not the cable or terminal). Touch the other probe of the DMM to the starter solenoid stud (again, not the cable itself). While holding the probes in this position, crank the engine until you get a steady reading. Ideally, there should be less than a one-half volt drop. If the drop is greater than one-half volt, narrow the problem down by checking point to point along the cable (e.g., battery post to cable end, cable end to cable wire, wire end to wire end). Follow the same procedure noted earlier in Task A.3. Ideally, each connection along the cable should have less than one-tenth of a volt drop. Remember that polarity with a DMM is not important. If the leads are backward, the display will show a minus sign in front of the reading.

Another way to check for high resistance is to crank the engine for several seconds and then feel along all the connection points and conductors for heat. If any point is more than just warm, there could be excessive voltage drop at that location. Clean or replace as required.

10. Inspect, test, and replace starter control circuit switches, relays, connectors, terminals, and wires (including thermal over crank protection).

The starter control circuit includes those components between the key start switch and the magnetic switch (or starter solenoid itself, if there is no magnetic switch in the circuit). It is important to remember that this is a critical, high-current circuit. Current requirements of solenoids on some larger starters are upward of 100 amps, often necessitating a separate relay (also known as a magnetic switch) to handle the high amperage load. Magnetic switch (also known as MAG) assemblies are typically four-post units with two large and two small terminals. However, some can also resemble large 5-pin mini-relays.

The key start switch is a low-current switching device that controls the magnetic switch (relay). Current from this relay goes to the starter solenoid (another relay) to engage the starter. Both the control and light current circuits need to be in working order to get the engine to crank. Also, neutral safety switches are sometimes found between the key switch and the magnetic relay. They are designed to prevent an engine from starting in gear.

Some manufacturers do not use a separate magnetic switch between the key switch and the starter solenoid, which makes the key switch circuit moderately high current. With a neutral safety switch and some bulkhead harness connections also part of the system, plenty of potential exists for trouble in the form of a no-start condition if everything is not in good working order. It is best to consult the wiring diagram for the vehicle to effectively troubleshoot the starting control circuit. Properly understanding relays and voltage drop testing is essential for diagnosing starting circuit complaints.

Some starter manufacturers incorporate a built-in circuit breaker that protects the starter from thermal damage. This circuit typically resets when the starter temperature drops below a preset temperature.

11. Diagnose starter cranking inhibit systems; determine needed repairs.

Some trucks use a cranking system inhibitor function to prevent the truck from cranking under certain conditions. Trucks with automatic transmissions use a switch on the transmission as an input device for this system. Starter operation is prevented when the truck is in gear: The starter will only function when the gear selector is in park or neutral. The technician can troubleshoot this system by checking the starter relay circuit. The park/neutral switch is often located in the relay coil ground circuit. If the starter relay coil lacks a ground, then find the park/neutral switch and inspect it closely. The switch is a very simple "open/closed" style switch that can be tested with an ohmmeter. A faulty switch should be replaced.

12. Inspect, test, and replace starter relays and solenoids/ switches including integrated MAG switch (IMS).

Most starter solenoids have two functions: They shift the starter drive pinion into mesh with the flywheel ring gear and make the high-current connection to deliver battery power to the starter motor field coils.

If a starter fails to crank and all the other circuits to the solenoid check out, it is important to remember that the set of electrical contacts in the solenoid are subject to pitting and corrosion over time. Testing them is easy. Perform a voltage drop test across the two large posts on the solenoid while cranking the engine, just as for a battery cable or clamp.

If a starter pinion fails to engage the ring gear, the cause could be a solenoid fault. When replacing a solenoid, keep in mind that quite often a pinion clearance adjustment needs to be made at the same time. See Task B.10 for more information on MAG switches.

13. Inspect, clean, repair, or replace cranking circuit cables, connectors, and terminals.

Test starter circuit components using the voltage drop method discussed in Task B.9. This is the preferred method for determining the serviceability of cranking circuit components. Obvious physical damage, such as cables with rubbed insulation or clamps with broken or cracked ends, warrants replacement.

If the battery cable is OK but needs new terminals, avoid using generic bolt-on cable terminals. They do not have the same current carrying capability as crimp or solder-type terminals and will almost certainly cause trouble later with corrosion and high resistance. Use the appropriate size crimp or soldered terminal, along with shrink tubing, when repairing battery cables to minimize corrosion and resistance.

14. Remove and replace starter; inspect flywheel ring gear or flex plate.

When removing a starter motor, inspect the teeth on both the starter drive pinion and the flywheel ring gear. Compare them with new components if unsure of what is worn. Keep in mind that both the pinion and ring gear have a machined chamfer on the teeth to facilitate engagement.

If it is determined that the flywheel ring gear needs to be replaced, see the engine manual for the appropriate procedure. If the starter pinion is worn, it can sometimes be replaced economically, compared to the cost of replacing the whole starter. However, take into consideration the age and condition of the starter motor itself before attempting a pinion drive replacement.

Vehicles with automatic transmissions use a flex plate to connect the engine crankshaft to the transmission torque converter. This flex plate has a ring gear on the outer edge that performs the same function as the flywheel ring gear. This flex plate and ring gear need to be closely inspected when servicing the starter or any of the transmission components in this area. If the technician finds cracks in the flex plate, then the flex plate should be replaced. The ring gear should be replaced if the teeth are worn or damaged.

15. Differentiate between electrical and/or mechanical problems that cause a slow crank, no crank, extended cranking, or a cranking noise condition.

There are many causes of slow crank, no crank, extended cranking, or cranking noise conditions; determining whether the cause is related to the engine or the electrical cranking circuit can be challenging to the technician. When such problems occur, the technician must evaluate whatever evidence is apparent by listening, checking that electrical connections are not overheating, and observing smoke emission during cranking that would indicate that fuel was being injected. The guided diagnostic software used by some engine manufacturers is designed to isolate electrical, electronic, and mechanical problems in a systematic manner. Use it when it is available. In most cases this software is designed to rapidly source electrical and fuel system problems, and can additionally direct the technician to locate engine mechanical problems.

Manually barring an engine will quickly verify whether an engine is mechanically seized. To manually bar an engine, the technician physically connects a large wrench to the engine. This process allows the engine to be turned by hand.

C. Charging System Diagnosis and Repair (7 questions)

1. Verify operation of charging system circuit monitors; determine needed repairs.

When a dash meter indicates a charging system problem, it is important to understand how that instrument is connected into the system. Many late-model trucks use an electronic dash display managed by a microprocessor. These systems can be scanned with a scan tool that allows the technician to retrieve trouble codes and check the data list, as well as calibrate the gauges.

Charge indicator lights generally illuminate with the key on and the engine off, and should extinguish when the engine starts. Any other action indicates a possible problem with the charging system. Be sure to properly troubleshoot the system before condemning an alternator as the source of the problem.

Ammeters are gauges that indicate how much amperage is either going into or leaving the battery. They are not often used by manufacturers, as a small current loss from the battery in an undercharging situation would not be detectable by most ammeters.

Voltmeters are a more reliable indicator of charging system condition, along with the state of charge of the battery. If the voltmeter reads over 13.5 volts, the alternator can be assumed to be charging. If it maintains that reading under the heaviest electrical loads, the alternator has been effectively load tested. If battery voltage drops below 13.5 volts under any conditions with the engine running at high idle, the alternator is undercharging. Conversely, a reading exceeding specification means the system is overcharging. Charging voltage should not exceed 14.2 volts in electrical systems using gel cell batteries.

2. Diagnose the cause of no charge, low charge, or overcharge conditions; determine needed repairs.

A no charge complaint could be caused by any number of reasons, from a simple blown fuse to faulty brushes or diodes inside the alternator. Due to the many different types of systems, it is impossible to list all the potential causes. Consult the service manual for the system you are working on and follow the troubleshooting procedures.

A low charge condition is one in which the alternator charges properly with light loads, but falters under heavy ones. See Task C.4 for details on diagnosing this complaint.

An overcharge condition (system operating over the maximum specified voltage) usually indicates a defective voltage regulator. In some systems, a defective sensing diode in the alternator can send a low signal to the regulator, forcing it to overcompensate.

Some manufacturers now use charging systems that allow the engine computer to control charging output. The technician should become familiar with the design of the electrical system prior to attempting to troubleshoot a charging problem. The technician should also perform voltage drop tests on the charging circuit each time an alternator is being diagnosed. See Task C.5 for details about how to perform this test.

See the diagnostic table below for further information on charging problems.

Condition	Possible Cause
Undercharging	■ Loose generator belt
	■ Faulty voltage regulator
	■ Faulty generator
	■ Poor connections at generator or battery
	■ Faulty battery
No Charging	■ Loose or missing generator belt
	■ Faulty voltage regulator

(Continued)

Condition	Possible Cause
	■ Faulty generator
	■ Poor connections at generator or battery
	■ Faulty battery
Overcharging	■ Faulty voltage regulator
	■ Faulty generator
	■ Poor connections at generator or battery
	■ Faulty battery

3. Inspect, adjust, and replace alternator drive belts/gears, pulleys, fans, mounting brackets, and tensioners.

Worn drive belts are visually obvious, but the technician should try to determine the cause of the failure. Two of the more common causes are incorrect belt tension and misalignment. On vehicles without automatic tensioners, recheck belt tension after installing a new belt. Allowing the system to run for a while will seat the new belt, after which readjustment is required. A loose belt will slip and fail prematurely.

Some alternators are shimmed fore and aft in their mounting brackets. Be sure to align the alternator-driven pulley with the engine drive pulley. An alternator that is not properly shimmed or aligned will cause premature belt wear, as well as a possible squeaking noise while the engine is running.

Many belt systems use a spring-loaded tensioner that needs to be checked for the correct spring tension with a large wrench or with a socket and a long ratchet. Alternators that still use an adjustable belt need to be adjusted manually to the correct tightness and then secured.

4. Perform charging system voltage (AC and DC) and amperage output tests; determine needed repairs.

Alternator output tests are simple procedures. All that is required is a battery load tester and a voltmeter. Before starting, look up the model ID on the alternator and check the manufacturer's specifications for maximum output. Also, check the drive belt for proper tension.

Attach the battery load tester across the battery terminals as if you were going to load test it. For this test, however, attach the amp clamp (either the one with the load tester or a handheld meter) around the alternator output wire. With the engine at high idle, load the battery tester down to a value slightly higher than the alternator's rated output. This forces the alternator to output maximum amperage, which should be within five percent of specs. Note that with the load tester drawing more current than the alternator can replace, the system voltage will be down. This is normal. Next, reduce the current draw on the load tester to an amount slightly less than the alternator's rated output. System voltage should then increase to over 13.5 volts. The technician should also measure the amount of AC voltage being produced by the charging system during these tests. There should not be more than 0.5 volts AC coming from the alternator. If the AC voltage is excessive, the diodes in the alternator are defective and will need to be serviced.

Some systems can be tested for maximum output with a procedure known as full-fielding. This is not recommended because unregulated voltage can damage electrical and electronic components.

5. Perform charging circuit voltage drop tests; determine needed repairs.

For the alternator to provide maximum output, the output wire and grounding of the alternator must be in good condition. Check both by testing voltage drop in the positive and negative sides of the circuit.

Voltage drop in the charging circuit should be performed while the truck is charging at a high rate. This test can be completed by first performing the maximum output test (Task C.4) and then measuring voltage drop between the output stud on the alternator and the positive battery cable. The reading should be less than 0.5 volts. Anything more indicates high resistance between the two points. With the alternator at maximum output, test for voltage drop between the alternator housing and the battery ground terminal. The result of the ground-side test should be less than 0.5 volts. If the positive side or ground side of the charging circuit has a voltage drop greater than 0.5 volts, the technician should investigate the cause. Items to look for include loose connections, burned connections, damaged wires, or wires that are too small.

6. Remove and replace alternator.

Replacing an alternator is a simple job. Label wires before removing them. Be sure to disconnect the negative battery cable before removing the alternator wiring and after reinstalling the new alternator to prevent accidental sparks and possible wiring damage. The alternator wires need to be installed securely to assure proper operation. Properly torque the fasteners and ensure the pulleys are aligned.

7. Inspect, repair, or replace charging circuit connectors, terminals, and wires.

In order for the alternator to transfer all of its power efficiently to the battery and the various loads in the vehicle, the wiring attached to the alternator must have clean connections and be in good condition. Other than a visual inspection for chafes, corrosion, etc., the best way to test the output wire is with the voltage drop test discussed in Task C.5.

When repairing wiring in the charging system, use the proper size cabling and connectors for the circuit you are working on. Charts are available that list the recommended wire gauge size depending on amperage flow and length of wire. Also, use the proper crimp tools along with shrink tubing to ensure a good connection.

D. Lighting Systems Diagnosis and Repair (6 questions)

1. Diagnose the cause of brighter than normal, intermittent, dim, or no headlight and daytime running light (DRL) operation; determine needed repairs.

An alternator that is overcharging is the only cause for multiple lights that are brighter than normal. Verify this by checking alternator output with a DMM. A malfunctioning charging system can also cause dim light operation if chassis voltage is too low. Verify this problem with a DMM, also.

Dim lights can also be caused by problems such as excessive resistance in fuse holders, relays, wiring, switches, connectors, and chassis grounds. Of all the aforementioned possibilities, suspect the chassis grounds first. Poor grounds cause the majority of problems related to this complaint, from simple loose hardware to poor metal-to-metal contact between chassis components. Voltage drop tests (see Task A.1) between the ground side of the bulb and the battery negative post will confirm ground integrity problems.

Intermittent operation can be caused by a cycling circuit breaker or a loose connection somewhere in the circuit. Circuit breakers are often incorporated directly into the headlight switch. A cycling circuit breaker is caused by either a defective breaker or an overload in the system, and will result in on-off circuit operation. Loose connections are harder to find. A good tip would be to gently pull and wiggle suspect harnesses and connectors while watching the light action.

A blown fuse, defective circuit breaker, bad switch, or an open in the wiring generally causes no light operation. At this point, it is best to get a wiring diagram for the vehicle and probe with a test light at various points downstream from the power source until you find the open.

Daylight running lights (DRL) are now mandated in some jurisdictions in the United States and in all of Canada. Some OEMs specify DRLs as a default option and fleets are using them increasingly. DRLs are a safety feature and illuminate anytime the vehicle is running and the vehicle headlights are not switched on. The DRL circuit is illuminated at key-on when the vehicle is in gear.

See the table below for more information on troubleshooting lighting problems.

Condition	Possible Cause
Brighter than normal	■ High charging system voltage
	■ Dimmer switch in high beam position
	■ Flash to pass or high beam relay stuck on
	■ Bulb problems
	■ Incorrect setting of automatic dimming sensor
Intermittent operation	■ Connection issues with switches, fire wall bulkhead connectors, in-line connectors, head lamp connectors, fuse, or relay blocks
	■ Weak circuit breakers
	■ Problems with head lamp relays
	■ Shorts causing circuit breaker shutdown
	■ Bulb problems
	■ Faulty relays
	■ Faulty switches (headlight or dimmer)
Dim	■ Poor grounds
	■ Contact corrosion in head lamp relays
	■ High resistance in connections at switches, fire wall bulkhead connectors, in-line connection, head lamp connectors, fuse, or relay blocks
	■ Bulb problems

(Continued)

Condition	Possible Cause
No headlights	■ Bulb problems
	■ Open or high resistance connection issues with switches, fire wall bulkhead connectors, in-line connectors, head lamp connectors, fuse, or relay blocks
	■ Damaged circuit breakers or fusible links
	■ Faulty relays
	■ Faulty switches (headlight or dimmer)
	■ Shorts causing circuit breaker shutdown

2. Inspect, replace, and aim headlights and auxiliary lights.

Headlight aim should be checked on a level floor with the vehicle unloaded. In some states, this may conflict with existing laws and regulations. If so, modify the instructions to meet the state's legal requirements.

To adjust headlights, first check headlight aim. Various types of headlight aiming equipment are available commercially. When using aiming equipment, follow the instructions provided by the equipment manufacturer.

When headlight aiming equipment is not available, aiming can be checked by projecting the upper beam of each light upon a screen or a chart at a distance of 25 feet ahead of the headlights.

Some manufacturers recommend coating the prongs and base of a new sealed beam with dielectric grease for corrosion protection. Use an electrical lubricant approved by the manufacturer.

Sealed-beam halogen and xenon headlights are designed to give substantially more light on high beam than are incandescent bulbs, extending the driver's range of visibility for safer night driving. Both halogen and xenon bulbs produce a whiter light, which helps improve visibility. They also last longer, stay brighter, and use less wattage for the same amount of light produced. When replacing individual replaceable bulbs, avoid touching the glass envelope. Oil from the skin can cause the bulb to shatter when turned on. Many auxiliary lights have replaceable bulbs and care should be taken when replacing these bulbs.

3. Inspect, test, repair, or replace headlight switches, dimmer switches, control components, relays, sockets, connectors, terminals, and wires.

Headlight switches are used to control the operation of the headlights and sometimes the parking and dash lights as well. Some manufacturers use separate toggle or rocker switches for each function, while others incorporate all switching functions into one multi-function switch. It is important to understand how a particular system is constructed to make troubleshooting easier. Always consult the wiring schematic for the vehicle you are working on.

Dimmer switches switch current flow from either the low- or high-beam circuit. This switch can either be mounted on the floor or incorporated into a multi-function turn signal switch, sometimes known as a stalk switch. The dimmer switch is a simple device that directs voltage to either the high or low beams.

Many headlight circuits contain relays to reduce the load conducted through the switch itself. A defective headlight or dimmer switch should affect both left and right headlights because they are wired in parallel. If only the high-beam or only the low-beam circuits are not operational, however, the headlight switch can usually be ruled out, because it is the dimmer switch that distributes power to the lights.

4. Inspect, test, repair, or replace truck and trailer lighting circuit switches, bulbs, light-emitting diodes (LEDs), sockets, control components, relays, connectors, terminals, and wires.

Parking, clearance, and taillight circuits can be controlled by a multi-function headlight switch or be powered by separate toggle or rocker switches. When troubleshooting these lights, keep in mind that the most common causes of trouble are usually related to poor grounds, either at the lights themselves or at a ground strap. Trailer connector plugs and sockets can be a source of problems when the pins corrode.

If only one clearance, taillight, or parking light is dim and the rest are OK, suspect a problem with that particular light or its ground. If all of the lights are dim, then assume that a ground or power supply malfunction is the cause.

Light-emitting diode (LED) lighting units are increasingly used on trucks and trailers. Despite higher initial costs, LEDs produce no heat and last much longer with less maintenance. Use OEM instructions for testing, but bear in mind that these are diodes, so polarity is important when testing them.

5. Inspect, test, repair, or replace instrumentation light circuit switches, bulbs, LEDs, sockets, fiber optic cable, circuit boards, connectors, terminals, wires.

Dash illumination lights (not to be confused with warning lights) on most trucks are fed from the same power source as the vehicle's taillights and clearance lights, usually through the multi-function light switch assembly. Between this source and the lights themselves, however, there is usually a rheostat, otherwise known as a dimmer switch, which allows the driver to reduce the intensity of the lights when driving at night. This dimmer switch may either be incorporated into the headlight switch or be remotely mounted elsewhere on the dash.

If the dash lights on a truck are dimmer than normal and all the other lights work OK, suspect a problem with the dimmer switch or its wiring, assuming that it has been adjusted to the correct position.

It may be sometimes necessary to replace the bulbs, LEDs, or fiber optic cable in the instrument panel area. The technician should exercise care when making repairs in this area due to all of the wires and fittings. In addition, fiber optic cable is composed of fragile glass fibers that can be broken if handled roughly.

6. Inspect, test, repair, or replace interior cab light circuit switches, bulbs, LEDs, sockets, connectors, terminals, and wires.

When diagnosing problems related to interior cab lights, the technician should remember that the door switches are on the ground side of the circuit. This enables the manufacturer to use a single wire to each switch. A short to ground on the switch side of the lamp will cause the lights to remain on constantly and not blow the fuse.

7. Inspect, test, adjust, repair, or replace stoplight circuit switches, relays, bulbs, LEDs, sockets, connectors, terminals, and wires.

Stoplight circuits are relatively simple in that the major component is the switch itself. There are two types. On medium-duty trucks with hydraulic brakes, the switch is usually located on the brake pedal, activated by simple mechanical movement. On larger trucks with air brakes, the switch is incorporated into one of the brake application air lines. Air pressure acting against a diaphragm will close electrical contacts in the switch to complete the circuit. Some trucks will have two such switches: one for service brake applications and another for parking brake applications.

Stoplight and turn signal lamps share the same bulbs, making it necessary to route the current from the stoplight switch through the turn signal switch. This allows it to be directed properly when both the brakes and turn signals are activated at the same time.

8. Diagnose the cause of turn signal and hazard light system malfunctions; determine needed repairs.

The turn signal circuit directs output from the flasher unit to one side of the truck or the other, depending on the position of the switch. If one side works properly but the other does not, assume that the flasher unit is OK and that the problem lies in the malfunctioning circuit. If the problem exists in the trailer and not the truck, a good place to start troubleshooting would be at the trailer connector plug. Each circuit can be separated and diagnosed at this point.

The hazard lights use the same bulbs, connectors, and wiring as the turn signals. The hazard lights use a separate flasher assembly. If the hazard lights become inoperative, the technician needs to check the turn signal operation. If the turn signals work correctly, then the likely problem would be the flasher.

The brake lamp switch can also be considered part of the turn signal circuit because, if the brakes are applied during a turn, the direction in which the truck is being turned must continue to flash. This is accomplished by routing the current from the stoplight switch directly into the turn signal switch. At this point, the turn signal switch will direct the current flow in the proper direction depending on position.

9. Inspect, test, repair, or replace turn signal and hazard circuit flashers or control components, switches, relays, bulbs, LEDs, sockets, connectors, terminals, and wires.

Turn signal flashers that operate faster or slower than normal can indicate a problem in the circuit. For example, if a bulb burns out on one side of a vehicle, this would reduce current demand in that particular circuit, usually causing the flasher assembly to blink more slowly than normal. Conversely, if a condition causes higher than normal current flow in the circuit, as when extra lights are added, current flow would be increased, causing the bi-metal contacts inside the flasher to heat up and cycle at a faster rate. This condition could cause premature failure of the flasher assembly.

Late-model vehicles with electronic flashers incorporate a relay in the circuit to handle the high-current switching demands and are less affected by defective or additional lights in the circuit. Electronic flashers are serviced in the same way as the older style flashers.

10. Inspect, test, adjust, repair, or replace back-up light and warning devices, circuit switches, bulbs, LEDs, sockets, connectors, terminals, and wires.

Switches mounted on the transmission or shift linkage usually control the back-up circuit. When the transmission is shifted into reverse, contacts inside the switch are closed and the back-up circuit is energized.

Audible and visual warning alarms are installed on some trucks, especially inner-city delivery vehicles, to alert persons near the truck that it is about to back up. These devices use the same switch and signal as the back-up lights, but also incorporate a relay in the circuit to handle the higher power demands of the alarms.

11. Inspect and test trailer light cord connector and cable; determine needed repairs.

The seven-wire trailer cord has become standard on U.S. highways and any technician working on highway trucks should be fully familiar with the wiring color codes used.

- White — Dedicated ground
- Black — Clearance, side marker, and identification lights
- Yellow — Left-hand turn signal and hazard signal
- Red — Stoplights (in some systems: ABS power)
- Green — Right-hand turn signal and hazard signal
- Brown — Tail and license plate lights
- Blue — Auxiliary or ABS power/power line carrier (PLC)

The power line carrier (PLC) is the channel used by the trailer ABS electronics to signal trailer ABS malfunctions to the driver dash display. PLC transmits a radio frequency (RF) signal over the ABS power line.

E. Related Vehicle Systems Diagnosis and Repair (12 questions)

1. Diagnose the cause of intermittent, inaccurate, or no gauge readings; determine needed repairs.

When troubleshooting gauge systems, it is important to determine whether the problem exists with all the gauges or just one gauge. For example, older style bi-metal gauges are sensitive to voltage fluctuations. If you find that they all read too high or too low, it is possible that the instrument voltage regulator is malfunctioning. The purpose of the instrument voltage regulator is to maintain a steady voltage supply to the gauges, regardless of battery voltage. A number of factors, including high resistance, a defective sending unit, or a malfunctioning gauge, may cause individual gauges to read erratically.

Follow the manufacturer's instructions for out-of-dash gauge testing. A good place to begin diagnosis on a single gauge problem is to disconnect the sending unit. Doing this should cause the gauge to react either by moving to the top or bottom of the scale. Next, try connecting the sending unit wire to a good ground. This action should cause the cause to move to the opposite end of the range as before. Some technicians will use a variable resistor at the sending unit to make the gauge move throughout its range. If the gauge reacts to the above actions, then the gauge and related wiring is good and the sending unit should be replaced.

Late-model instrumentation systems primarily use magnetic-style gauges. Most are not affected by changes in system voltage and therefore do use a separate instrument voltage regulator. Follow OEM instructions when diagnosing electronic dash displays.

Not all gauges in an instrument panel are electrically operated. Some gauges, such as air application gauges, are mechanically actuated and not affected by electrical problems. Air application gauges must test within 4 psi of an accurate master gauge.

Never rely only on the accuracy of dash gauges to make major decisions, such as determining the need for an engine rebuild. For example, if vehicle instrumentation indicates low engine oil pressure, confirm the problem with a known, accurate, mechanical gauge to verify the condition.

2. Diagnose the cause of data bus driven gauge malfunctions; determine needed repairs.

Instrumentation systems in late-model vehicles rely on data collected from sensors that transmit information to many chassis control system modules. Each module can broadcast information over a data bus to the other on-board computers. One computer is usually dedicated to instrumentation. It is the function of the instrumentation module to receive digital signals from the chassis data bus and display them on the dash in analog, bar graph, or digital format, depending on vehicle options.

Newer instrumentation systems go into a self-test mode when the key is first turned on to verify the operation of each gauge. When the accuracy of any gauge is suspect, first determine if the dash is receiving accurate information. To do this, connect the appropriate electronic service tool (EST) to the vehicle diagnostic connector and scroll the menu options to scan the various sensor values that the module is receiving. From there you can determine the problem source by checking the values against actual vehicle operating parameters.

3. Inspect, test, adjust, repair, or replace gauge circuit sending units, sensors, gauges, connectors, terminals, and wires.

To test sending units, use the diagnostic electronics or service manual for the vehicle to determine what the senders should read at specific operating parameters. For example, a certain resistance value might be specified for a fuel level sending unit when the tank is half full. Or, a temperature sending unit could be tested in boiling water, again, with a specified resistance value at that point. Another quick way to check thermistor-type temperature senders on a vehicle with an electronic engine is to compare actual values (using a diagnostic tool) to those specified in the OEM software parameters.

Many electrical gauges and some electronic ones can be checked using a variable resistance test box. Unplug the wire at the sending unit and substitute the resistance of the test box while comparing the action of the gauge against manufacturer specifications.

Some gauges, such as pyrometers, do not use a separate power source. Their purpose is to measure exhaust gas temperature. Heat energy is used to generate a small voltage, which is then correlated with a temperature value at the dash display.

Replacing sending units and sensors are typically fairly simple operations. One caution to keep in mind is to allow the engine to cool down prior to replacing engine temperature sending units or engine temperature sensors. Replacing other types of sending units involves removing the connector and then physically removing it from its location before reinstalling the new component.

4. Inspect, test, repair, or replace warning devices (lights and audible) circuit sending units, sensors, circuit boards/control modules, bulbs, audible component, sockets, connectors, terminals, and wires.

Warning lights and buzzers on older, pre-electronic component vehicles have a sender or switch on the ground side of each monitored function that close and allow the circuit to be powered up in the event of a malfunction.

Late-model vehicles have built-in warning systems designed to alert the driver in case of a malfunction that might damage the engine or other components. Usually, an audible alarm unit is located behind the dash and will sound when a malfunction is detected. Alarm units are often incorporated into the dash itself and not serviced separately; however, some are remotely mounted and can be replaced.

Trucks with electronically controlled engines have at least two dash warning lights to indicate faults. One will be a "check engine" light, and the other a "stop engine" light. The former will alert the driver to a condition that needs attention at the driver's earliest convenience, while the latter will indicate a problem that requires immediate engine shutdown, such as low oil pressure or high water temperature. Depending on how the engine protection system has been programmed, the stop engine light alert may be accompanied by either a ramp-down in engine power and/or an automatic engine shutdown.

When testing circuits in an electronic instrumentation system, always use a DMM to check voltage: Never use test lights. Test lights draw too much current from an electronic circuit to make this type of testing valid and may damage the circuit. It should be noted that most electronic circuits operate at voltages lower than battery voltage.

5. Inspect, test, and replace electronic instrumentation systems; verify proper calibration for vehicle application.

Most speedometers, odometers, and tachometers on late-model trucks are controlled electronically by means of information received from the data bus (see Task E.2). When the information on any of these indicators is suspect, first check the calibration menu of the manufacturer diagnostic software tool to see how the vehicle speed parameters are set up. For example, in order for the computer to accurately display speed or mileage data, it must first know the rear axle ratio and tire size. Some engine/vehicle management systems even track tire wear to adjust the speedometer reading. If these parameters are not properly programmed into the computer, inaccurate data will result. These data fields may be password protected. The manufacturer puts passwords on some types of parameters to prevent unauthorized changes to settings.

Many speedometers and tachometers obtain their information through sensors known as magnetic pickups. These are simple devices that create a magnetic field near a rotating, toothed wheel. As the wheel rotates and the teeth cut the magnetic field, a small voltage signal is generated in the pickup. The ECM uses signal frequency to calculate engine or wheel speed.

Some older vehicles with electric gauges source the tachometer signal off the alternator, usually the "R" terminal. Use a signal generator for an easy way to test these gauges if suspect. This device simulates the signal that would normally be fed from a magnetic pickup or alternator phase tap. Consult the manufacturer's repair manual to correlate the frequency of the signal being sent to the gauge with the actual reading.

6. Diagnose the cause of constant, intermittent, or no horn operation; determine needed repairs.

An electric horn uses high current so the circuit almost always includes a relay. If the horn does not work at all, first check for power going to the horn itself. If the unit is properly grounded and does not operate, the unit is defective and should be replaced. Check both the control and power circuits at the relay.

7. Inspect, test, repair, or replace horn circuit relays, horns, switches, clock spring, connectors, terminals, and wires.

When testing a horn relay circuit, first check for fused power at the positive side of the horn relay (usually pin #30). Next, check for a signal from the steering wheel switch. This may be either positive or ground-side switching, depending on the manufacturer. Assuming the use of a 5-pin mini-relay, the circuit should function exactly as described in Task A.8.

8. Diagnose the cause of constant, intermittent, or no wiper operation; diagnose the cause of wiper speed control and/ or park problems; determine needed repairs.

Most electric windshield wiper systems use multi-speed motors. The motor is either a permanent magnet type with high- and low-speed brushes, or one that uses an external resistor pack, much like heater blower motors. Inside the first style of motor is a low-speed and a high-speed brush set. If the motor works OK in one speed, but is sluggish or not functioning in the other, it may be due to defective brushes inside the motor. First, ensure that the proper voltage is coming from the appropriate side of the wiper switch with the circuit closed. Consult the manufacturer's repair manual for the relevant vehicle wiring diagram.

Intermittent wiper systems use the same basic components as a regular wiper with the addition of a logic device. This logic device can be located on the wiper motor or within a body control module. This device has a timing function incorporated to make the wipers delay their wiping pattern, based upon where the driver has the switch set. The technician needs to use a high impedance multimeter when testing the circuit.

9. Inspect, test, and replace wiper motor and transmission mechanical linkage, arms, and blades, relays, switches, connectors, terminals, and wires.

If a wiper system parks consistently in the wrong position, it may be due to a misadjusted wiper crank arm linkage. Readjust the linkage, using factory indexing marks where applicable. Binding linkage may cause wiper action that is sluggish or intermittent. Disconnect the motor crank arm and manually move the wiper linkage to be sure that the mechanisms are free to operate without undue force. Wipers that fail to park properly after the switch is turned off may be due to a defective park switch, which is usually incorporated into the motor itself. The function of this switch is to continue motor operation until the blades are fully retracted. Also, check that the park switch is receiving power when the wiper switch is in the "off" position.

Replacing the wiper motor and/or the transmission linkage usually requires removal of the cowl panel. The technician should pay close attention to the timing of the motor to the transmission to assure proper operation.

10. Inspect, test, repair, or replace windshield washer motor or pump/relay assembly, switches, connectors, terminals, and wires.

Most washer pump circuits are relatively simple. The switch for the pump is usually incorporated into the wiper control switch and will control the washer pump through either positive or ground-side switching. If the wipers work but the washers do not, you can eliminate a fuse or circuit breaker problem since the wiper switch assembly usually draws power from the same source. After checking to make sure that there is washer fluid in the reservoir, the technician should use the process of elimination to troubleshoot an inoperative washer. Listen to the washer pump motor for noise when the switch is depressed. If the pump does make a sound, then check the hoses and nozzles for a restriction. If the pump does not produce any sound, then the technician should make sure the motor is getting voltage and is connected to a good ground. If voltage and a good ground are present, then the motor should be replaced.

11. Inspect, test, repair, or replace sideview mirror motors, heater circuit grids, relays, switches, connectors, terminals, and wires.

Moveable mirror assemblies incorporate two different motors, one for horizontal movement and the other for vertical. If one or both motors do not operate, first confirm that power exists at the motor terminals when the appropriate side of the mirror switch is depressed.

Heated mirrors can be considered a safety option in cold climates. When testing a heater circuit for proper operation, be aware that some circuits contain a fast warm-up cycle, which then drops considerably in current draw as the mirror warms.

12. Inspect, test, repair, or replace heater and A/C electrical components including: A/C clutches, motors, resistors, sensors, relays, switches, control modules, connectors, terminals, and wires.

Heater blower motors typically have three or four speeds. They work by routing current flow from the switch through a set of resistors wired in series. If one of these resistors should fail, the motor may work in high speed only, since the resistors are by-passed in the high-speed position. Blower motor resistors are usually found in the blower airstream to keep them from overheating. Even though this voltage drop is intentional, it still produces heat.

A/C compressor clutches can be controlled by either a set of manual switches and pressure sensors, or directly from an A/C system microprocessor. Research how the system is configured before troubleshooting. See Tasks A.10, A.11, and A.12 for information on troubleshooting electronically controlled systems.

A/C compressor clutches may need to be shimmed properly during their replacement to maintain proper air clearance in the disengaged position. This must be done carefully. Too little clearance may cause the clutch to drag and burn up; too much may prevent engagement.

13. Inspect, test, repair, or replace cigarette lighter and/or auxiliary power outlet, integral fuse, connectors, terminals, and wires.

Many vehicles today have what are known as auxiliary power outlets in place of cigarette lighters. These are convenience items for drivers should they need to plug in portable accessories, such as a laptop computer or navigational aid (e.g., global positioning system or GPS device).

Some of these outlets look like cigarette lighter sockets, while others are multiple-pin connectors designed to provide both key switch-operated (on-off) and constant power. When troubleshooting these devices, remember that they consist of at least two different circuits.

14. Diagnose the cause of slow, intermittent, or no power window operation; determine needed repairs.

A power window that operates sluggishly or not at all may be due to either incorrect voltage being supplied to the motor or a binding window mechanism. To differentiate between the two, check for the proper voltage at the motor when the switch is actuated. If the voltage is correct and/or the system continually blows fuses, binding linkage is most likely the cause. When troubleshooting these devices, remember that both the power and the ground side of the circuit should be checked. A power window that operates intermittently can be caused by faulty brushes in the motor or by a wiring connection problem. The technician should attempt to investigate what causes the motor to start working. If the motor works after hitting the door panel with a rubber hammer, then the motor is likely to have worn brushes, which would necessitate replacing the motor. If the motor starts to work after manipulating the wiring and connectors, then there is likely a poor connection that will need to be repaired.

15. Inspect, test, repair, and replace power window motors, switches, relays, connectors, terminals, and wires.

If both windows in a truck are inoperative, the power source is a good place to start troubleshooting. If one side works but the other does not, check the switch, followed by testing voltage at the motor.

When testing a power window motor circuit, keep in mind that the motors are reversible. This means that the wiring connected to the motor does not have a dedicated power and ground assignment. Rather, the assignment alternates at the control switch, depending on the direction in which the switch is activated.

16. Diagnose inverter/shore power systems; determine needed repairs.

Many trucks that have sleepers incorporate a system that allows the operation of creature comfort features, such as the air conditioner, heater, microwave, and refrigerator, without the truck engine running. These systems incorporate an inverter that converts 12 volt DC energy into 110 volt AC energy. This system is self-supporting as long as the batteries do not discharge too much. If the batteries discharge too much, then the truck automatically starts and the alternator charges the batteries. Some systems also have a shore power option that lets the driver plug into a port at a truck stop that is equipped with shore power cords. The driver can run all of the AC-powered items at the same time that the batteries are getting charged up while plugged in at a truck stop. These systems greatly reduce fuel consumption and emissions.

Auxiliary power units (APUs) are used to reduce engine idle periods in many late-model trucks. An APU is an electric power generating system designed to provide for the chassis electrical requirements when not running on the highway. APU systems can be powered by a small internal combustion engine that is connected to the truck fuel system with lines and hoses. Some manufacturers use battery-powered APU systems.

17. Diagnose the cause of poor, intermittent, or no operation of electric door locks; determine needed repairs.

The locking and latching mechanisms of door locks should be lubricated with dry graphite because it is not sticky and will not attract dust and debris. Truck door locks increasingly use auxiliary electrical controls that can be actuated by hard-wired switches or wirelessly. It should be easy to determine whether a problem is electrical or mechanical by observation and listening. Eliminate the possibility of a mechanical problem before focusing on the door lock's electrical circuit. Use the OEM wiring schematic to search for the causes of electrical malfunctions and exercise care when disassembling the door assembly.

18. Inspect, test, repair, or replace electric door lock circuit switches, relays, controllers, actuators/solenoids, connectors, terminals, and wires.

Most late-model trucks use auxiliary electrical control of door lock circuits. OEMs typically provide an isolated schematic that makes it easier to read. The electrical hardware used consists of a low-current control circuit connected to a relay that switches the lock solenoids. Both the relay and solenoid produce an audible click when actuated, which can help in troubleshooting. Performing a door lock repair or replacement procedure will normally involve removing the door panel. This process involves removing the bolts and clips that hold the door panel in place. The technician should carefully account for all of the hardware and strive to reinstall all of these items when completing the repair.

19. Inspect, test, and replace cruise control electrical components.

Cruise control in all late-model trucks is managed by the ECM, chassis electronics, or both off the vehicle data bus (J1939). Critical inputs to the cruise control system are information from the vehicle speed sensor located in the transmission tail shaft, chassis brake status, and the specific speed value programmed into the management system by the driver. There are two general categories of cruise control. The first uses a hard value. Hard value cruise control attempts to permit no upper or lower droop over the programmed road speed, regardless of the operating/road conditions encountered. This can significantly lower vehicle fuel efficiency.

Most late-model trucks are equipped with "soft" or "smart" cruise control. Depending on the system, soft cruise does not close-loop actual road speed based on the driver-programmed value. Instead, the system will permit both upper- and lower-speed droop to occur based on such factors as the terrain the vehicle is traveling through and the percentage of engine load. For instance, in the interest of vehicle fuel economy, it makes sense to allow road speed to drop slightly going uphill and increase slightly going downhill. Soft cruise may also do such things as engage the engine fan and use the engine compression brake circuit to maintain a programmed road speed value.

When reprogramming, troubleshooting, or repairing electronic cruise control systems, the OEM service software and data connection hardware must be used. If you have no access to the OEM diagnostic software, refer the unit to a dealer.

20. Inspect, test, and replace engine cooling fan electrical control components.

Most fans on highway trucks are air-operated devices that obtain their air signal from a 12 volt solenoid. This solenoid is actuated by some sort of temperature-sensing switch on older vehicles or by the engine computer on late-model vehicles. Verify which type you are working on before troubleshooting.

On systems managed by a remote temperature switch, the switch will either control the power or ground side of the circuit. When it makes the internal connection, an air solenoid is energized, and air is directed or exhausted to either engage or disengage the fan, depending on the type of fan.

On ECM-controlled fan circuits, it is important to understand the control system logic for the engine/vehicle you are working on. In addition to the normal on-off operation described earlier, some control systems are programmed to engage the fan whenever a temperature sensor on the engine fails. This is a fail-safe protection feature for the engine, when the module can no longer determine the actual temperature. Other systems engage the fan whenever any portion of the circuit fails by reverse ground-side switching. Again, this is a fail-safe feature designed to protect the engine. Trucks may also engage the fan during cruise control set speed overrun (downhill driving) conditions to assist in vehicle retardation.

An air-operated fan clutch may use air to apply or release the fan, depending on the model of fan. Both types use an engine coolant electrical sensor assembly to control compressed airflow to the fan hub through an electrically actuated air supply-exhaust valve solenoid. The system is available as either a normally closed (NC) or normally open (NO) electrical control system. These systems also have an optional manual toggle control switch. Battery power is supplied to the optional control switch and to the engine thermal time switch from the ignition feed. When the engine coolant temperature reaches a preset value of the thermal time switch, an electrical circuit is completed to the NO or NC solenoid fan clutch valve. A drop in engine temperature will cause the NO or NC thermal time switch to break the electrical circuit, which de-energizes the fan clutch hub. Trucks with A/C systems will also use a refrigerant pressure switch. The refrigerant pressure switch is a NC two-pole switch that opens when A/C compressor head pressure exceeds 300 psi. When the head pressure returns to less than 210 psi, the switch will close.

Replacing the engine cooling fan electrical control components is typically very simple. The electrical connector needs to be removed and then the component can be physically removed from the truck. Care should be taken when reinstalling the component and electrical connection to assure a quality repair.

21. Inspect, test, and replace electric fuel supply/transfer pump control components.

Fuel supply/transfer pumps are used to pull fuel from the fuel tanks up to the engine. These electric fuel pumps are typically energized by a relay that is controlled by the engine computer. A scan tool can be used to check the engine computer for codes related to the fuel pump operation. Some manufacturers also will have an output test option available on the scan tool. This test will send a signal to the pump to turn on. Like any electric motor, the fuel pump needs to have the correct voltage supplied to it in order to function correctly. A voltage supply test will reveal any voltage supply problems in the circuit. The fuel pump should receive at least 12 to 13 volts. If the supplied voltage is less than specified, then the power and ground circuit should be checked. Voltage drop tests can be performed while the circuit is energized to locate wiring and connection problems. If the correct amount of voltage is supplied to the pump but it is still inoperative, then the pump should be replaced with a quality replacement unit. The fuel system is a very delicate system that necessitates servicing fuel filters on a regular basis. Follow the manufacturer's recommendations when servicing the fuel filters on each truck.

Sample Preparation Exams

INTRODUCTION

Included in this section is a series of six individual preparation exams that you can use to help determine your overall readiness to successfully pass the Electrical/Electronic Systems (T6) ASE certification exam. Located in Section 7 of this book you will find blank answer sheet forms you can use to designate your answers to each of the preparation exams. Using these blank forms will allow you to attempt each of the six individual exams multiple times without risk of viewing your prior responses.

Upon completion of each preparation exam, you can determine your exam score using the answer keys and explanations located in Section 6 of this book. Included in the explanation for each question is the specific task area being assessed by that individual question. This additional reference information may prove useful if you need to refer back to the task list located in Section 4 of this book for additional support.

PREPARATION EXAM 1

1. A resistance test was performed on an open toggle switch and the result is 5.857 megohms. Technician A says that the switch is faulty because the reading is beyond the specifications. Technician B says that the switch has over 5 million ohms of resistance. Who is correct?

    A. A only
    B. B only
    C. Both A and B
    D. Neither A nor B

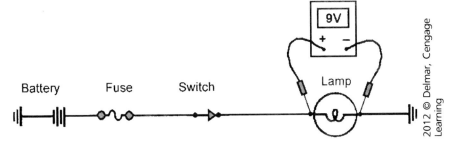

2. The 12 volt battery in the figure is fully charged and the switch is closed. Which of the following conditions would be the most likely cause of this measurement?

    A. Larger than specified wire used in wire repair
    B. Loose ground for the bulb
    C. Blown fuse
    D. Faulty bulb

3. Which of the following conditions could cause the power locks to fail to operate?

    A.   Broken connector at the power lock switch
    B.   Closed power lock circuit breaker
    C.   An oversized fuse
    D.   An oversized wire repair in the power lock circuit

4. When using a 12 volt, unpowered test lamp to test for voltage, all of the following are true EXCEPT:

    A.   The test lamp may be connected to the battery ground.
    B.   The battery should first be disconnected.
    C.   The test lamp may be connected to a chassis ground.
    D.   The test lamp will not illuminate when probed into the circuit at a point after the load.

5. After removing a 30 amp circuit breaker from a fuse panel, a technician checks continuity across its terminals with a digital multimeter (DMM). Technician A says that the current from the ohmmeter will open the circuit breaker. Technician B says the ohmmeter should read infinity if the circuit breaker is good. Who is correct?

    A.   A only
    B.   B only
    C.   Both A and B
    D.   Neither A nor B

6. Technician A uses a test lamp to detect resistance. Technician B uses a jumper wire when verifying the operation of circuit breakers, relays, and switches. Who is correct?

    A.   A only
    B.   B only
    C.   Both A and B
    D.   Neither A nor B

7. Technician A says that a charge indicator light should come on with the key on, engine off. Technician B states that if the indicator is not on with the engine running, it can be assumed that the charging system is functioning properly. Who is correct?

    A.   A only
    B.   B only
    C.   Both A and B
    D.   Neither A nor B

8. Technician A says that an amp clamp is a useful tool to use when checking for key-off drain. Technician B says that a faulty rectifier bridge in the alternator could cause excessive key-off drain. Who is correct?

    A.   A only
    B.   B only
    C.   Both A and B
    D.   Neither A nor B

9. All of these conditions would cause the starter not to crank the engine EXCEPT:

   A. The battery is not connected to the starter motor.

   B. The solenoid does not engage the starter drive pinion with the engine flywheel.

   C. The control circuit fails to switch the large-current circuit.

   D. The starter drive pinion fails to disengage from the flywheel.

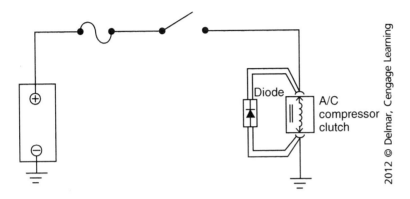

10. Referring to the figure above, what is the diode being used for?

    A. Allow the A/C compressor to run in one direction only.

    B. Protect the A/C compressor clutch from a voltage spike.

    C. Protect the circuit from a possible voltage spike.

    D. Limit the current flow to the A/C compressor clutch.

11. The stoplights are inoperative on a truck equipped with air brakes. Technician A says that a faulty pressure switch could be the cause. Technician B says that a faulty hazard lamp switch could be the cause. Who is correct?

    A. A only

    B. B only

    C. Both A and B

    D. Neither A nor B

12. Technician A says that a solenoid has a movable metallic plunger that creates linear movement when the coil is energized. Technician B says that some solenoids are used to control fluid in hydraulic circuits. Who is correct?

    A. A only

    B. B only

    C. Both A and B

    D. Neither A nor B

13. The charging light stays on while driving. Technician A says that a grounded wire near the alternator could be the cause. Technician B says that a faulty circuit in the instrument cluster could be the cause. Who is correct?

    A. A only

    B. B only

    C. Both A and B

    D. Neither A nor B

14. All of the following electrical tools could be used to diagnose a data bus network problem EXCEPT:

    A.   Oscilloscope
    B.   Digital multimeter
    C.   Analog voltmeter
    D.   Scan tool

15. How many load test amps are pulled from the battery during a high rate discharge (load) test?

    A.   Two times the cold cranking ampere (CCA) task rating of the battery
    B.   One and one-half times the CCA rating of the battery
    C.   One-half the CCA rating of the battery
    D.   Three-fourths the CCA rating of the battery

16. A customer has a truck towed to the shop and says that the starter would not crank the engine. What should be checked first?

    A.   Ground cable connection
    B.   Starter solenoid circuit
    C.   Ignition switch crank circuit
    D.   Battery for proper charge

17. An alternator is being replaced on a heavy-duty truck. Technician A uses an air tool to remove the fastening nut from the charging output wire. Technician B disconnects the negative battery cable prior to removing the alternator. Who is correct?

    A.   A only
    B.   B only
    C.   Both A and B
    D.   Neither A nor B

18. Technician A says that batteries can be recharged more quickly by using a high setting on the battery charger. Technician B says that batteries can be charged more thoroughly by using a low setting on the battery charger. Who is correct?

    A.   A only
    B.   B only
    C.   Both A and B
    D.   Neither A nor B

19. A heavy-duty truck is slow to crank. Technician A says that testing the voltage drop on the positive battery cable while cranking the engine could reveal potential problems in the battery. Technician B says that testing the voltage drop on the negative battery cable must be done with the ignition switch in the off position. Who is correct?

    A.   A only
    B.   B only
    C.   Both A and B
    D.   Neither A nor B

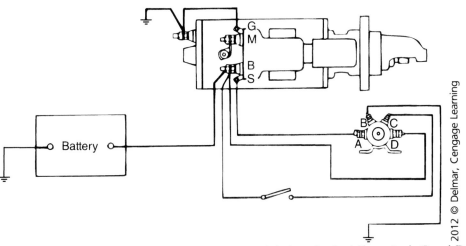

20. Referring to the figure above, which component is being checked if terminals C and D are jumped at the magnetic switch?

  A. Starting switch
  B. Battery
  C. Magnetic switch
  D. Starter

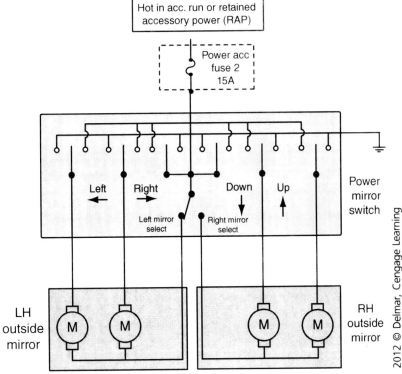

21. Referring to the figure above, the left mirror functions normally, but the right-side power mirror does not function at all. Technician A says that mirror select switch could be defective. Technician B says that the power mirror switch could have a disconnected ground. Who is correct?

  A. A only
  B. B only
  C. Both A and B
  D. Neither A nor B

22. Which of the following steps should the technician do first when beginning to remove a starter from a truck?

    A. Pull the starter fuse at the power distribution center.
    B. Remove the starter bolts.
    C. Disconnect the wires at the starter.
    D. Remove the negative battery cable and tape the terminal.

23. A technician is working on a truck and finds that a battery cable terminal end is badly corroded. All of the following are proper repair procedures EXCEPT:

    A. Replace the entire cable assembly.
    B. Replace the terminal with a bolt-on end and heat-shrink tubing.
    C. Replace the terminal with a crimp-on end and heat-shrink tubing.
    D. Replace the terminal with a soldered end and heat-shrink tubing.

24. While checking the fuses in a tractor/trailer, the technician finds an open fuse. Which of the following is the next step?

    A. Replace the fuse with the next higher amperage rating.
    B. Check the affected circuit for a short to ground.
    C. Check the affected circuit for an open.
    D. Install a circuit breaker with a smaller amperage rating than the fuse.

25. Which of the following faults could cause a truck with an automatic transmission to have an inoperative cranking system?

    A. Faulty throttle pedal
    B. Blocked fuel injector
    C. Battery with terminal voltage of 12.6 volts
    D. Open wire at the park/neutral switch

26. Which of the following conditions would be the LEAST LIKELY cause for a blown horn fuse?

    A. Shorted electric horn
    B. Power wire rubbing a metal bracket
    C. Open terminal
    D. Shorted fuse box

27. A maintenance-free battery is low on electrolyte. Technician A says a defective voltage regulator may cause this problem. Technician B says a loose alternator belt may cause this problem. Who is correct?

    A. A only
    B. B only
    C. Both A and B
    D. Neither A nor B

28. Which of the following conditions could cause a misaligned alternator pulley?

    A. A/C compressor bearing
    B. Loose alternator mounting bracket
    C. Loose alternator output wire
    D. Water pump bearing

29. What is the LEAST LIKELY result of a full-fielded alternator?

    A. Low battery voltage due to excessive field current draw

    B. Burned-out light bulbs on the vehicle

    C. Battery boil over

    D. Excessive charging system voltage

30. A heavy-duty truck is being diagnosed for a charging problem. The alternator only charges at 12.2 volts. Technician A says that the voltage drop should be checked on the charging output wire. Technician B says that the voltage drop should be checked on the charging ground circuit. Who is correct?

    A. A only

    B. B only

    C. Both A and B

    D. Neither A nor B

31. What is the LEAST LIKELY cause of a discharged battery?

    A. Loose alternator belt

    B. Corroded battery cable connection

    C. Defective starter drive

    D. Parasitic drain

32. Which of the following conditions would most likely cause a burned charging output wire?

    A. Failed alternator rotor assembly

    B. Incorrect alternator pulley

    C. Loose alternator bracket

    D. Voltage regulator stuck at maximum charging mode

33. The right-side headlight on a truck is very dim, and the left-side headlight is normal. Technician A says that the dimmer switch is likely defective. Technician B says that the left-side headlight could have a bad ground. Who is correct?

    A. A only

    B. B only

    C. Both A and B

    D. Neither A nor B

34. Technician A says that headlight aim should be checked on a level floor with the vehicle unloaded. Technician B says that some states have very strict laws on adjusting the headlights. Who is correct?

    A. A only

    B. B only

    C. Both A and B

    D. Neither A nor B

35. The dash lights on a medium-duty truck do not work. Technician A says that the fuse to the taillights could be the cause. Technician B says that the rheostat in the headlight switch could be the cause. Who is correct?

    A. A only

    B. B only

    C. Both A and B

    D. Neither A nor B

36. Technician A says that all International Organization for Standardization (ISO) relays are the same and can be interchanged. Technician B says that terminals 85 and 86 of an ISO relay are connected to the coil inside the relay. Who is correct?
    A. A only
    B. B only
    C. Both A and B
    D. Neither A nor B

37. The right turn signal indicator comes on when the brakes are applied. Technician A says that a bad ground at the right rear stop lamp socket could be the cause. Technician B says that a single-filament bulb in a dual-filament socket could be the cause. Who is correct?
    A. A only
    B. B only
    C. Both A and B
    D. Neither A nor B

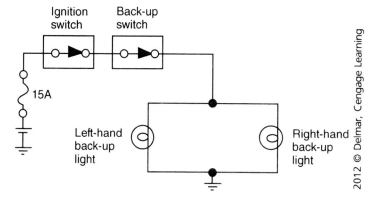

38. Referring to the figure above, the right-side back-up light circuit is accidentally grounded on the switch side of the bulb in the circuit. Technician A says this condition may blow the back-up light fuse. Technician B says the left-side back-up light may work normally while the right-side back-up light is inoperative. Who is correct?
    A. A only
    B. B only
    C. Both A and B
    D. Neither A nor B

39. Technician A says that a truck with an electronic instrument panel sources its data directly from the sensors on the engine. Technician B states that a data bus-style electronic instrument panel receives information exclusively from the engine control module (ECM). Who is correct?
    A. A only
    B. B only
    C. Both A and B
    D. Neither A nor B

40. The temperature sending unit needs to be replaced on a heavy-duty truck. Technician A says that this process can be completed without draining the whole cooling system. Technician B says that the engine should be cooled down prior to performing this repair. Who is correct?
    A. A only
    B. B only
    C. Both A and B
    D. Neither A nor B

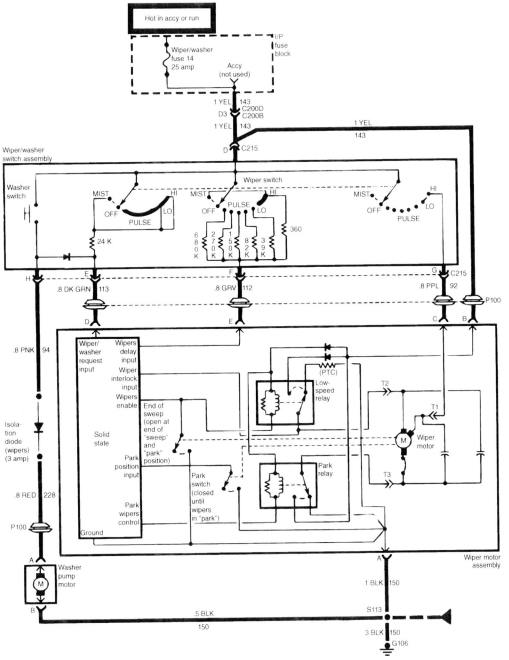

2012 © Delmar, Cengage Learning

41. Referring to the wiring schematic above, Technician A says that the wiper motor receives ground from item P100. Technician B says that Fuse #14 supplies power to the windshield wiper switch. Who is correct?

   A. A only
   B. B only
   C. Both A and B
   D. Neither A nor B

42. Which of the following components provides the tachometer signal to the instrument cluster on older medium-duty trucks with diesel engines?

    A. Primary ignition coil
    B. Secondary ignition coil
    C. Alternator (R terminal)
    D. Vehicle speed sensor

43. A customer says the electric horn on a truck will not turn off. Technician A says the cause could be welded diaphragm contacts inside the horn. Technician B says the relay may be defective. Who is correct?

    A. A only
    B. B only
    C. Both A and B
    D. Neither A nor B

44. The windshield washer pump motor runs continuously while the ignition switch is on. Technician A says that the multi-function switch could be shorted. Technician B says that the wiper control module could be defective. Who is correct?

    A. A only
    B. B only
    C. Both A and B
    D. Neither A nor B

45. Technician A says that the starter solenoid pull-in winding can be tested by touching the "start" terminal and the "bat" terminal with the ohmmeter leads. Technician B says that touching the "start" terminal and the solenoid case with the ohmmeter leads will test the starter solenoid hold-in winding. Who is correct?

    A. A only
    B. B only
    C. Both A and B
    D. Neither A nor B

46. A blower motor will not work on high but works well on all other speeds. Technician A says that a faulty blower resistor could be the cause. Technician B says that a faulty blower relay could be the cause. Who is correct?

    A. A only
    B. B only
    C. Both A and B
    D. Neither A nor B

47. The auxiliary power outlet is inoperative and the fuse is found to be open. What is the most likely cause for this condition?

    A. Loose connection at the power outlet plug
    B. Foreign metal object in the power outlet
    C. Broken wire leading to the power outlet
    D. Open internal connection at the power outlet

48. Which of the following conditions would be the LEAST LIKELY cause for all of the power window motors to be inoperative?

    A. Open internal circuit breaker at the passenger power window motor
    B. Faulty power window master switch
    C. Missing ground at the master switch
    D. Faulty power window circuit breaker

49. Which of the following is a measurement for electrical pressure?

    A. Ohm
    B. Amp
    C. Watt
    D. Volt

50. A truck is being diagnosed for a fuel supply problem. The technician measures the voltage available at the transfer pump terminals and finds 7.5 volts. Technician A says that the problem could be caused by a faulty fuel pump relay. Technician B says that the problem could be caused by a poor ground connection in the supply pump circuit. Who is correct?

    A. A only
    B. B only
    C. Both A and B
    D. Neither A nor B

PREPARATION EXAM 2

1. Technician A says that burned electrical contacts will decrease the electrical resistance in a circuit. Technician B says that an open switch should have continuity. Who is correct?

    A. A only
    B. B only
    C. Both A and B
    D. Neither A nor B

2. A truck is in the repair shop with an inoperative power window. During diagnosis, a blown power window fuse is located. Technician A says that a short to ground between the switch and the motor could be the cause. Technician B says that a tight power window motor could be the cause. Who is correct?

    A. A only
    B. B only
    C. Both A and B
    D. Neither A nor B

3. A truck is being diagnosed for a problem with the cruise control kicking out when the turn signal is operated. Which of the following would be the most likely cause?

    A. Faulty throttle pedal
    B. Faulty multi-function switch
    C. Faulty clutch switch
    D. Faulty engine control module (ECM)

4. Technician A says that a digital multimeter (DMM) can be used to test current flow directly through the meter in any truck electrical circuit. Technician B states that a current clamp can be used on high-amperage circuits to prevent damage to the meter. Who is correct?

    A. A only

    B. B only

    C. Both A and B

    D. Neither A nor B

5. The blower motor in a truck is running very slowly. An ammeter shows a low current flow. Which of the following could cause the described conditions?

    A. High resistance in the circuit

    B. Low resistance in the circuit

    C. Overcharged battery

    D. Shorted blower motor

6. Technician A says that loose electrical contacts will decrease the electrical resistance in a circuit. Technician B says that an open switch should read very low when using an ohmmeter. Who is correct?

    A. A only

    B. B only

    C. Both A and B

    D. Neither A nor B

7. A truck is being diagnosed for a charge indicator light that is not illuminated when the engine is running. The truck has an undercharged battery. Technician A says a blown fuse between the indicator lamp and ignition switch could cause this problem. Technician B says this could be caused by a failed charge light bulb. Who is correct?

    A. A only

    B. B only

    C. Both A and B

    D. Neither A nor B

8. Technician A says that a key-off current draw of two amps could cause the batteries to discharge. Technician B says that two amps is a normal key-off draw for a vehicle with an ECM. Who is correct?

    A. A only

    B. B only

    C. Both A and B

    D. Neither A nor B

9. A truck is in the repair shop for a starting problem. There is a clicking sound at the starter when the key is moved to the start position. Technician A says that the starter solenoid could be the cause. Technician B says that worn starter brushes could be the problem. Who is correct?

    A. A only

    B. B only

    C. Both A and B

    D. Neither A nor B

10. Which of the following could result if an air conditioner (A/C) compressor clutch diode fails "open"?

    A. ECM failure from voltage spikes
    B. Compressor running backward
    C. Clutch coil inoperative due to no current
    D. Clutch coil failure from high current

11. Which of the following problems would most likely cause the stoplights to be inoperative?

    A. Stuck closed lamp back-up switch
    B. Incorrect turn signal flasher
    C. Blown headlight fuse
    D. Blown stoplight fuse

12. A relay can be tested with a multimeter for all of the following tests EXCEPT:

    A. Resistance of the normally open (NO) contacts
    B. Resistance of the clamping diode
    C. Resistance of the normally closed (NC) contacts
    D. Resistance of the coil

13. Which of the following generally activates warning lights and/or warning devices?

    A. Vehicle ignition switch
    B. Closing a switch or sensor
    C. Opening a switch or sensor
    D. Vehicle battery

14. Which of the following tools is the LEAST LIKELY choice for use in diagnosing a problem on the data bus network?

    A. Oscilloscope
    B. DMM
    C. Continuity tester
    D. Scan tool

15. Technician A says that the battery must be at least 40 percent charged in order for load test to be valid. Technician B says that the load test must be run for 10 seconds. Who is correct?

    A. A only
    B. B only
    C. Both A and B
    D. Neither A nor B

16. Which of the following would be the most likely terminal voltage on a fully charged truck battery?

    A. 12 volts
    B. 12.2 volts
    C. 12.6 volts
    D. 13.2 volts

17. An alternator is being replaced on a heavy truck. Technician A uses an air tool to install the fastening nut onto the charging output wire. Technician B carefully routes the drive belt around all of the pulleys before releasing the belt tensioner. Who is correct?

    A. A only

    B. B only

    C. Both A and B

    D. Neither A nor B

18. Technician A says that a low battery cannot generate explosive vapors to the extent that a fully charged battery can. Technician B states that it is good practice to wear eye protection when jump-starting a truck. Who is correct?

    A. A only

    B. B only

    C. Both A and B

    D. Neither A nor B

19. A truck is being diagnosed for an inoperative starter. A voltage drop test is performed on the solenoid "load side" while the ignition switch is held in the crank mode and 0.1 volts are measured. Technician A says that the solenoid is faulty. Technician B says that this test should only be performed if the battery is at least 75 percent charged. Who is correct?

    A. A only

    B. B only

    C. Both A and B

    D. Neither A nor B

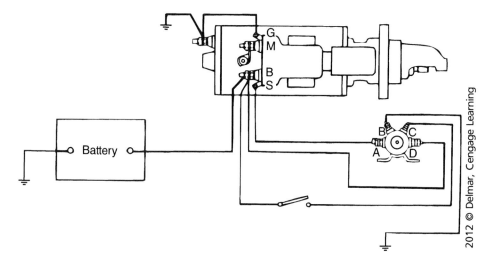

20. Referring to the figure above, which component is being checked if terminals A and D are jumped at the magnetic switch?

    A. Starting switch

    B. Battery

    C. Magnetic switch

    D. Starter

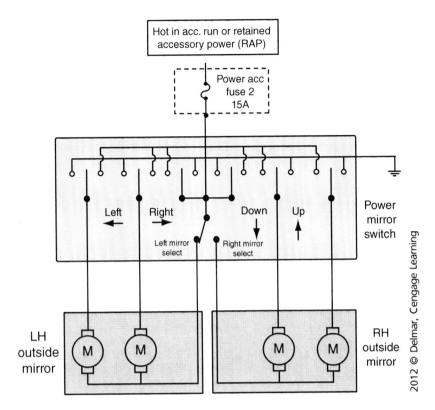

21. Referring to the figure above, the right-side power mirror functions normally but the left-side power mirror does not function in the up and down directions. Technician A says the mirror select switch could be defective. Technician B says the built-in circuit breaker in the up/down motor could be defective. Who is correct?

    A. A only
    B. B only
    C. Both A and B
    D. Neither A nor B

22. Technician A says that a starter drive pinion should not have chamfers on the drive teeth. Technician B states that if the flywheel ring gear is damaged, then the entire flywheel should be replaced. Who is correct?

    A. A only
    B. B only
    C. Both A and B
    D. Neither A nor B

23. Technician A says that the battery cables only need to be serviced when the starting or charging system is producing problems. Technician B says that battery corrosion only forms on the terminals during cold weather. Who is correct?

    A. A only
    B. B only
    C. Both A and B
    D. Neither A nor B

24. Which of the following statements describes a maxi-fuse?

    A. Higher-quality fuse than standard
    B. Higher-capacity fuse than standard and used in place of fusible links
    C. Another name for a circuit breaker
    D. Found in all truck electrical circuits because of their high current requirements

25. Technician A says that a scan tool can be used to retrieve trouble codes from an on-board computer of a truck. Technician B says that a multimeter can be used to retrieve trouble codes from a computer of a truck. Who is correct?

    A. A only
    B. B only
    C. Both A and B
    D. Neither A nor B

26. Technician A says an ammeter should be used to check for a short circuit between circuits. Technician B says to fully charge the battery before checking a circuit for current draw. Who is correct?

    A. A only
    B. B only
    C. Both A and B
    D. Neither A nor B

27. A truck has a problem of an alternator with zero output. Technician A says the alternator field circuit may have an open circuit. Technician B says the fusible link may be open in the alternator to battery wire. Who is correct?

    A. A only
    B. B only
    C. Both A and B
    D. Neither A nor B

28. The accessory drive belt system should be inspected during regular intervals. Technician A says that a serpentine drive belt tensioner should snap back after releasing pressure on it. Technician B says that the drive belt should be replaced at the first sign of cracks on the back side of it. Who is correct?

    A. A only
    B. B only
    C. Both A and B
    D. Neither A nor B

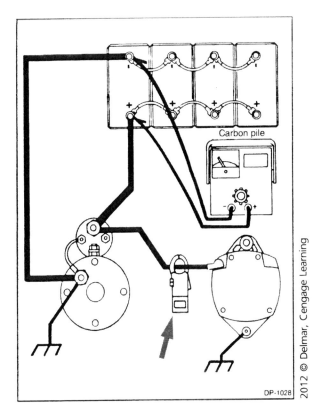

Carbon pile

DP-1028   2012 © Delmar, Cengage Learning

29. Referring to the figure above, what test is being performed with the instrument indicated by the arrow?

    A. Starter current draw test

    B. Battery load test

    C. Alternator output test

    D. Parasitic battery draw test

30. What is the most likely test that could prove the presence of a poor connection at the charging insulated circuit?

    A. Voltage drop test of the charging ground circuit

    B. Voltage drop test of the negative battery cable

    C. Voltage drop test of the positive battery cable

    D. Voltage drop test of the charging output wire

31. Technician A says that battery acid spills should be cleaned up when found in order to prevent major corrosion in the battery box. Technician B says that water and baking soda should be used to neutralize battery acid if spilled. Who is correct?

    A. A only

    B. B only

    C. Both A and B

    D. Neither A nor B

32. All of the following wire repair methods could be used on the charging circuit EXCEPT:

    A. Water-resistant harness replacement
    B. Crimp-style butt connectors
    C. Crimp-and-seal connectors
    D. Solder and heat shrink

33. All of the following switches could be incorporated into the multi-function switch EXCEPT:

    A. Dimmer switch
    B. Turn signal switch
    C. Stoplight switch
    D. Headlight switch

34. Technician A says that a corroded parking lamp socket could cause a brighter than normal lamp assembly. Technician B says that a single-filament bulb can be used in place of a dual-filament bulb. Who is correct?

    A. A only
    B. B only
    C. Both A and B
    D. Neither A nor B

35. Which of the following components could be used to progressively dim dash lights?

    A. Voltage limiter
    B. Rheostat
    C. Transistor
    D. Diode

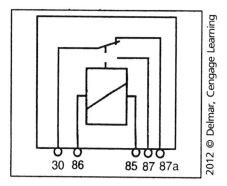

36. Referring to the figure above, each pin in a 5-pin mini-relay is identified with a number. What is pin #30 for?

    A. Control power in
    B. Control ground
    C. High-amperage power in
    D. High-amperage power out, normally closed

37. The back-up alarm is inoperative, but the back-up lamps work as designed. Technician A says that an open alarm relay could be the cause. Technician B says that a blown inline fuse to the alarm could be the cause. Who is correct?

    A. A only
    B. B only
    C. Both A and B
    D. Neither A nor B

38. The taillights on the tractor burn normally but the taillights on the trailer are dim. Technician A says that the light cord could have too much resistance. Technician B says that the tractor taillight bulbs could be open. Who is correct?

    A. A only
    B. B only
    C. Both A and B
    D. Neither A nor B

39. The oil pressure gauge intermittently moves to the area above the high setting. Technician A says that the oil pressure should be checked with a manual gauge to verify oil pressure. Technician B says that an open wire in the oil gauge could be the cause of this problem. Who is correct?

    A. A only
    B. B only
    C. Both A and B
    D. Neither A nor B

40. Which of the following devices is LEAST LIKELY to be used as an input to an electronic gauge assembly?

    A. Thermistor
    B. Piezo resistor
    C. Body control module (BCM)
    D. Light-emitting diode (LED)

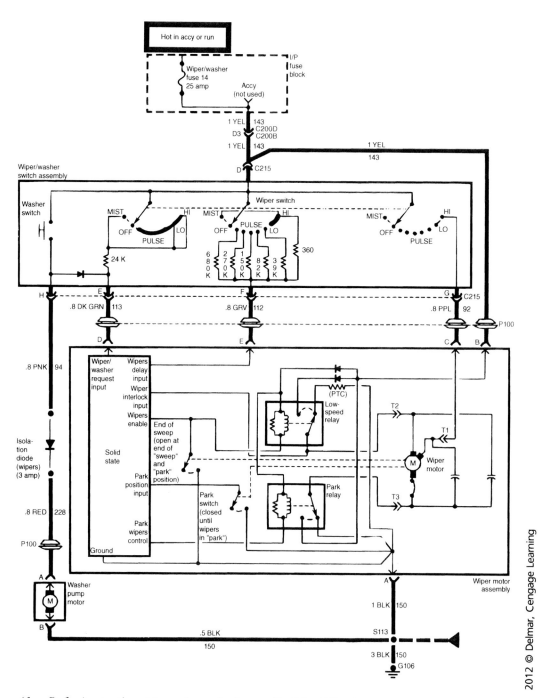

41. Referring to the wiring schematic above, all of the following statements are correct EXCEPT:

    A. The wiper switch is an input to the BCM.

    B. The windshield wiper circuit is grounded at G106.

    C. P100 is a pass-through grommet for the windshield wiper circuit.

    D. The wiper motor has two speeds.

42. The speedometer in an electronically managed truck is not accurate. Which of the following is the LEAST LIKELY problem?

    A. The rear axle ratio has not been correctly programmed to the engine computer.

    B. The transmission speed sensor has not been calibrated.

    C. The tire rolling radius has not been correctly programmed into the engine computer.

    D. The engine computer was not reprogrammed when new rear tires were installed.

43. A truck with dual electric horns is being diagnosed. Technician A says that the horns will likely be wired in series with each other. Technician B says that these systems typically use a high-note horn and a low-note horn. Who is correct?

    A. A only

    B. B only

    C. Both A and B

    D. Neither A nor B

44. Technician A says a binding mechanical wiper linkage can result in no wiper operation. Technician B says a control circuit shorted to ground can cause constant wiper operation. Who is correct?

    A. A only

    B. B only

    C. Both A and B

    D. Neither A nor B

45. Which of the following components is found in a truck starting circuit?

    A. Solenoid

    B. Ballast resistor

    C. Voltage regulator

    D. Engine control module (ECM)

46. Technician A states that auxiliary power outlet is powered at all times. Technician B states that an auxiliary power outlet can be used to jumpstart another truck. Who is correct?

    A. A only

    B. B only

    C. Both A and B

    D. Neither A nor B

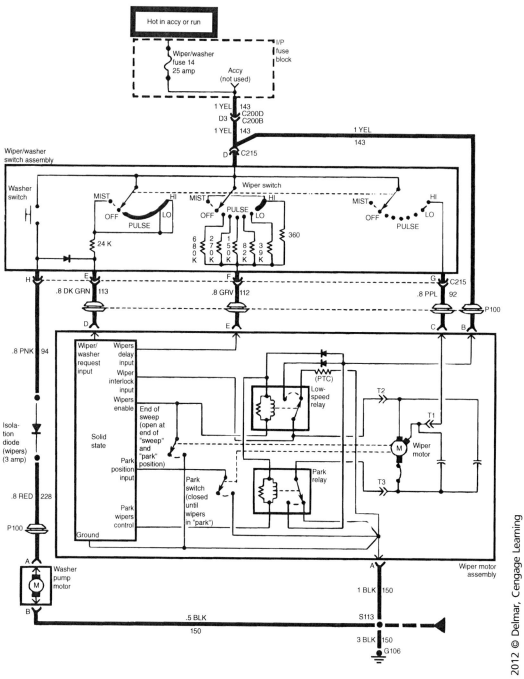

47. Referring to the wire diagram above, the windshield washer in the circuit shown does not
operate. The wiper motor operates normally. Technician A says the wiper/washer fuse may
be open. Technician B says the isolation diode may have an open circuit. Who is correct?

A. A only

B. B only

C. Both A and B

D. Neither A nor B

48. Which of the following steps would most likely be performed when replacing the passenger-side power door lock switch?

    A. Disconnect the clip from the lock switch.

    B. Remove the key lock tumbler.

    C. Remove the BCM.

    D. Remove the door panel.

49. When using a voltmeter to perform a voltage drop test in a circuit, the leads should be connected in what way?

    A. To the battery terminals

    B. From the positive battery terminal to ground

    C. In series with the circuit being tested

    D. In parallel with the circuit being tested

50. Which of the following engine parameters would LEAST LIKELY be present in the data list while using a scan tool?

    A. Throttle position

    B. Speed sensor resistance

    C. Turbo boost pressure

    D. Engine coolant temperature

PREPARATION EXAM 3

1. Technician A says that an electrical switch that has continuity will allow current to flow when the switch is closed. Technician B says that a piece of wire that has high resistance will have increased current flow. Who is correct?

    A. A only

    B. B only

    C. Both A and B

    D. Neither A nor B

2. Technician A says that a voltmeter could be used to find an open circuit by grounding the black lead and using the red lead to test for voltage available throughout the circuit. Technician B says that an open circuit would create a very high current flow and likely would blow a fuse. Who is correct?

    A. A only

    B. B only

    C. Both A and B

    D. Neither A nor B

3. Which of the following conditions could cause the power locks to operate intermittently?

    A. Blown power lock fuse

    B. Open power lock circuit breaker

    C. Broken wire near the power lock actuator

    D. Chaffed wire near the power lock switch

4. Technician A says that an open switch should have infinite resistance. Technician B says that a closed switch should have continuity. Who is correct?
   A. A only
   B. B only
   C. Both A and B
   D. Neither A nor B

5. A truck/trailer combination comes in with a dim left-rear taillight on the trailer. Technician A says that this may be due to a corroded pin on the left side of the trailer connector. Technician B states that the problem may be caused by a poor ground connection at the left-rear taillight. Who is correct?
   A. A only
   B. B only
   C. Both A and B
   D. Neither A nor B

6. Which tool would be LEAST LIKELY used when checking for key-off battery drain problems?
   A. Low amp probe
   B. Ohmmeter
   C. Ammeter
   D. Amp clamp

7. Which of the following is part of a truck charging system?
   A. Voltage solenoid
   B. Voltage regulator
   C. Voltage transducer
   D. Magnetic switch

8. Which of the following devices would most likely be used as a spike suppression device for an electromagnetic coil?
   A. Relay
   B. Diode
   C. Transistor
   D. Thermistor

9. A truck is being diagnosed for excessive starter noise while the engine is cranking. Technician A says that the starter may have incorrect clearance at the drive gear to ring gear. Technician B says that the starter mounting bolts may be overtightened. Who is correct?
   A. A only
   B. B only
   C. Both A and B
   D. Neither A nor B

10. What is the purpose of the clamping diode that connects in parallel with the coil of the relay?
    A. Controls voltage spikes as the relay is de-energized
    B. Controls voltage spikes as the relay is energized
    C. Assists in creating the magnetic field when the relay is energized
    D. Assists in creating the magnetic field when the relay is de-energized

11. Technician A says that all stoplight switches are air activated. Technician B states that stoplight switches route current directly to the stoplights. Who is correct?

A.   A only

B.   B only

C.   Both A and B

D.   Neither A nor B

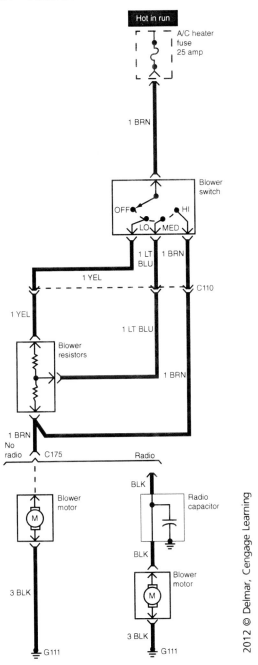

12. All of the following statements about the wiring schematic above are correct EXCEPT:

A.   The blower motor has three speeds.

B.   The blower switch by-passes the blower resistors when HI speed is selected.

C.   The A/C heater fuse provides power for the blower motor circuit.

D.   C175 provides ground for the blower motor.

13. The ground circuit for the fuel sender has failed. Technician A says that only the fuel gauge will be affected. Technician B states that all of the gauges will be affected because the gauges share a common ground. Who is correct?
    A. A only
    B. B only
    C. Both A and B
    D. Neither A nor B

14. Technician A says that the J1939 data bus uses two 120 ohm terminating resistors. Technician B says that the J1939 data bus uses a two-wire design. Who is correct?
    A. A only
    B. B only
    C. Both A and B
    D. Neither A nor B

15. All of the following are acceptable battery and cable maintenance procedures EXCEPT:
    A. Remove the negative battery cable last and reinstall first to avoid sparks.
    B. Clean corrosion and moisture accumulation on the battery top with a water and baking soda solution.
    C. Only replace battery cable ends with proper solder or crimp-on terminals and heat-shrink tubing.
    D. Coat battery terminal ends with a protective grease to retard corrosion.

16. A technician is attempting to charge a battery and yet, according to the ammeter on the charge, it will not accept a charge. What is the LEAST LIKELY source of the problem?
    A. The battery is already fully charged.
    B. The battery is highly sulfated.
    C. Poor contact exists between the charging clamp and the battery post.
    D. Excessive moisture accumulation has caused surface discharge.

17. Technician A says that some alternators are mounted using a pad-type bracket. Technician B says that some alternators are mounted using rivets. Who is correct?
    A. A only
    B. B only
    C. Both A and B
    D. Neither A nor B

18. The low-voltage disconnect (LVD) system opens (turns off power) when the battery voltage drops to what level?
    A. 8.4 volts
    B. 12.6 volts
    C. 10.4 volts
    D. 9.6 volts

19. Which of the following problems would be the LEAST LIKELY cause for an excessive voltage drop in the battery's positive cable?
    A. Corrosion in the battery cable
    B. Overcharged batteries
    C. Undersized positive battery cable
    D. Frayed wires in the battery cable

20. All of the following components are parts of the starter control circuit EXCEPT:

    A. Park/neutral switch
    B. Positive battery cable
    C. Starter relay
    D. Ignition switch

21. The wipers on a truck equipped with electric windshield wipers will not park. Technician A says the activation arm for the park switch is broken or out of adjustment. Technician B says a defective wiper switch will cause this condition. Who is correct?

    A. A only
    B. B only
    C. Both A and B
    D. Neither A nor B

22. Technician A says that the replacement starter assembly should be inspected carefully prior to its installation onto the engine. Technician B says that the replacement starter should be bench-tested prior to installing it onto the engine. Who is correct?

    A. A only
    B. B only
    C. Both A and B
    D. Neither A nor B

23. Technician A says that all replacement batteries should be fast charged for 15 minutes prior to their installation into the truck. Technician B says that the battery cable terminals should be cleaned and protected when installing a replacement battery. Who is correct?

    A. A only
    B. B only
    C. Both A and B
    D. Neither A nor B

24. Which of the following would most likely be the resistance test result of a good International Organization for Standardization (ISO) relay?

    A. 1 ohm when connected to terminals 30 and 85
    B. Out of limits (OL) when connected to terminals 30 and 87a
    C. 80 ohms when connected to terminals 85 and 86
    D. 1 ohm when connected to terminals 30 and 87

25. Technician A says that the main data link connector (DLC) is a round 9-pin connector. Technician B says that flash code diagnostics can be used to retrieve DTCs from the truck. Who is correct?

    A. A only
    B. B only
    C. Both A and B
    D. Neither A nor B

26. The function of a maxi-fuse is to:

    A. Take the place of a circuit breaker.

    B. Close during a current overload.

    C. Open and close when signaled by a computer.

    D. Open an overloaded circuit when excessive current is present in the circuit.

27. All of the following conditions could cause an undercharged battery EXCEPT:

    A. Loose alternator bracket fasteners

    B. An oversized charging wire

    C. Worn alternator brushes

    D. Excessive voltage drop in the charging wire

28. An alternator with a 90 ampere rating produces 45 amps during an output test. A V-belt drives the alternator and the belt is at the specified tension. Technician A says the V-belt may be worn and bottomed in the pulley. Technician B says the alternator pulley may be misaligned with the crankshaft pulley. Who is correct?

    A. A only

    B. B only

    C. Both A and B

    D. Neither A nor B

29. When performing an alternator maximum output test, what is the most practical and safe way to make the alternator produce maximum output?

    A. Place a carbon pile tester across the battery terminals.

    B. Full-field the alternator.

    C. Turn on all of the vehicle's electrical loads.

    D. Install a discharged battery into the vehicle.

30. Which of the following repair procedures would most likely correct an excessive voltage drop problem in the positive charging circuit?

    A. Tighten the starter solenoid attaching bolts.

    B. Replace the terminal at the alternator output wire.

    C. Tighten the alternator mounting bracket.

    D. Replace the voltage regulator.

31. A medium-duty truck with a dead battery is being jump-started. Technician A says the engine should be running on the boost vehicle before attempting to crank the dead vehicle. Technician B says the engine should be off while connecting the booster cables. Who is correct?

    A. A only

    B. B only

    C. Both A and B

    D. Neither A nor B

32. All of the following methods of wire repair in the charging system are currently used EXCEPT:

    A. Crimp-and-seal connectors

    B. Butt connectors and tape

    C. New wiring harness

    D. Solder and heat shrink

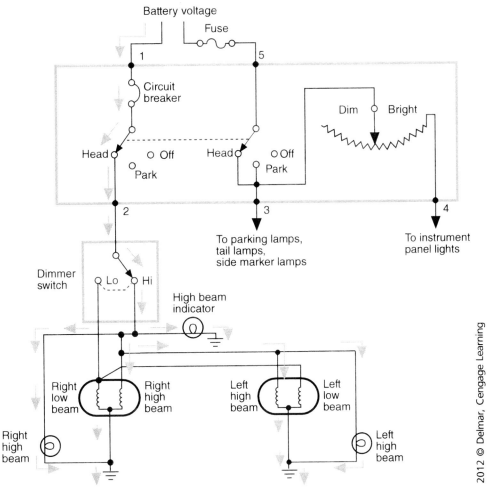

2012 © Delmar, Cengage Learning

33. The left-side headlight is dim only on the high beam in the figure above. The other headlights operate normally. Technician A says there may be high resistance in the left-side headlight ground. Technician B says there may be high resistance in the dimmer switch high-beam contacts. Who is correct?

    A. A only

    B. B only

    C. Both A and B

    D. Neither A nor B

34. A truck has an intermittent fault with its high beams only. All of these could be a possible cause EXCEPT:

    A. A defective headlight dimmer switch.

    B. Defective high-beam filaments inside headlamps.

    C. A loose wiring harness connector.

    D. A defective headlight switch.

35. A trailer has inoperative taillights on one side only. Technician A says to check the trailer circuit connector for an open. Technician B uses an ohmmeter to check the continuity between the defective side and the trailer circuit connector with the circuit under power. Who is correct?

    A. A only

    B. B only

    C. Both A and B

    D. Neither A nor B

36. A truck electrical system is being repaired. Technician A says that wiring schematics give exact details of the location of all electrical components on a truck. Technician B says that a wiring schematic will usually contain pin-out test procedures that can be used to troubleshoot many electrical faults. Who is correct?

    A. A only

    B. B only

    C. Both A and B

    D. Neither A nor B

37. If only the right side of the trailer's turn signals is illuminated, the technician should:

    A. Replace the turn signal flasher.

    B. Replace the bulbs on the left side of the trailer.

    C. Check the trailer electrical connection.

    D. Check the brake light switch for proper operation.

38. A trailer light cord power connector needs to be tested. Technician A says a digital multimeter (DMM) is an effective tool for testing the cord. Technician B says an incandescent test light is an effective method for testing the cord. Who is correct?

    A. A only

    B. B only

    C. Both A and B

    D. Neither A nor B

39. A heavy truck is being diagnosed for a fuel gauge that does not work. Technician A says that the other gauges should be checked for correct operation as part of the diagnosis process. Technician B says that the sending unit for the fuel gauge is in the fuel tank. Who is correct?

    A. A only

    B. B only

    C. Both A and B

    D. Neither A nor B

40. On a truck with electronic instrumentation, gauge accuracy is suspect. What would be the recommended method to determine where the fault lies?

    A. Swap the panel with a new one.

    B. Replace the suspect sender unit(s).

    C. Ground the sender wire at the suspect sender unit(s).

    D. Connect a scan tool and compare display to gauge readings.

41. A truck with a data bus network problem is being diagnosed. Technician A says that the driver will not likely notice any unusual problems if the data bus wires become shorted together. Technician B says data bus communication happens when voltage pulses are sent from module to module thousands of times per second. Who is correct?

    A. A only

    B. B only

    C. Both A and B

    D. Neither A nor B

42. The "check engine" light (CEL) illuminates while a truck is being operated. Which of the following causes would not require the driver's immediate attention?

    A. Low engine oil pressure

    B. High engine coolant temperature

    C. Maintenance reminder

    D. Low coolant level

43. A vehicle electric horn does not function when the horn is depressed. Technician A uses a test lamp to check the power and ground terminals of the horn relay. Technician B uses a DMM to check the power and ground terminals of the horn relay. Who is correct?

    A. A only

    B. B only

    C. Both A and B

    D. Neither A nor B

44. An older-type, two-speed wiper motor is being diagnosed. Technician A says that some trucks create two-speed operation by using separate sets of high- and low-speed brushes inside the motor. Technician B states that the two-speed operation is accomplished on some trucks by using an external resistor pack similar to a heater blower motor. Who is correct?

    A. A only

    B. B only

    C. Both A and B

    D. Neither A nor B

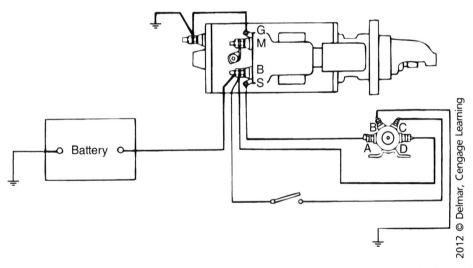

45. A technician is about to perform a voltage drop test across the starter solenoid internal contacts. Referring to the figure above, where should the voltmeter leads be placed?

    A. Between the positive battery terminal and point B

    B. Between the positive battery terminal and point M

    C. Between points B and M

    D. Between points G and ground

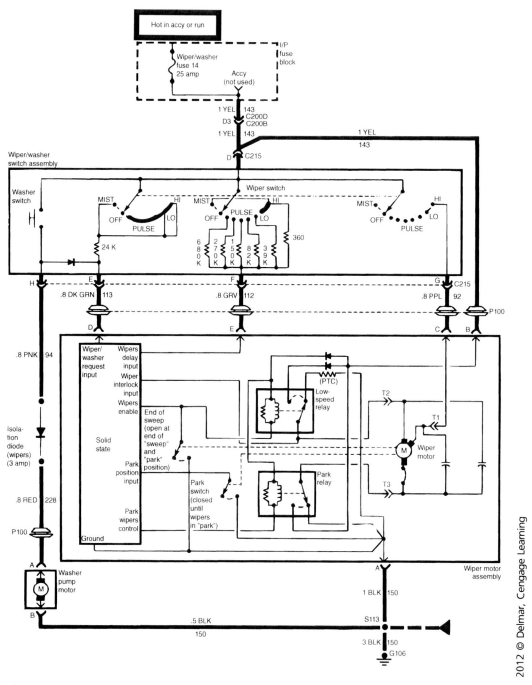

46. Referring to the figure above, the wiper washer pump motor runs constantly. Which of the following conditions would most likely cause this problem?

   A. The ground side of the motor is shorted to ground.

   B. The control switch is shorted to ground.

   C. The contacts in the switch are stuck closed.

   D. The wiper washer pump relay contacts are stuck closed.

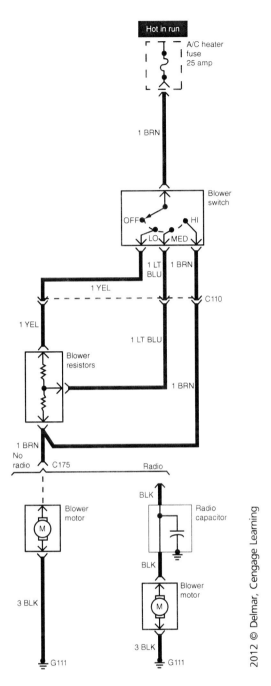

47. Referring to the figure above, when the blower resistors are removed, the blower motor will:

   A. Not operate at all.

   B. Blow the system fuse.

   C. Operate on high speed only.

   D. Operate on low speed only.

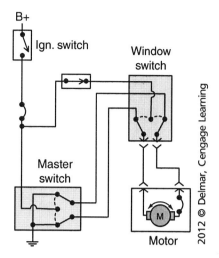

48. Referring to the figure above, the power window will not operate from either the master or window switches. Which of the following would be the most likely cause of this problem?

   A. Open diode

   B. Bad ground at the master switch

   C. Closed circuit breaker at the motor

   D. Shorted diode

49. Technician A says an ammeter is used to test continuity. Technician B says an ammeter measures current flow in a circuit. Who is correct?

   A. A only

   B. B only

   C. Both A and B

   D. Neither A nor B

50. Where would the power door lock relay most likely be located on the truck?

   A. Under the driver's seat

   B. In the under-hood power distribution center

   C. Near the rear heating, ventilating, and air conditioning (HVAC) assembly

   D. In the in-cab power distribution center

PREPARATION EXAM 4

1. Technician A says that a self-powered test lamp can be used to check continuity in a circuit managed by an electronic control module (ECM). Technician B states that an analog multimeter may be used to test voltages in an electronic circuit. Who is correct?

   A. A only

   B. B only

   C. Both A and B

   D. Neither A nor B

2. All of the following could be used when checking a circuit for continuity EXCEPT:
   A. An ohmmeter
   B. An ammeter
   C. A voltmeter
   D. A self-powered test light

3. A truck's electronic cruise control will not operate, although everything else functions properly. Which of the following should a technician check first?
   A. Servo motor
   B. Diagnostic software
   C. Linkage to the fuel pump
   D. Vehicle speed sensor

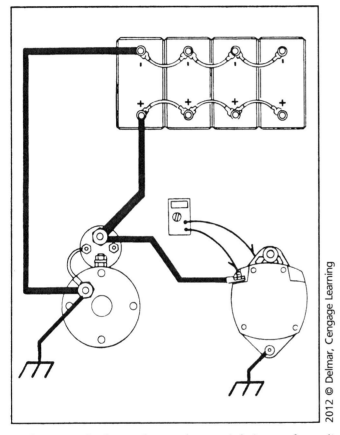

4. Referring to the figure above, what test is being performed?
   A. Voltage output test
   B. Positive charging circuit cable voltage drop test
   C. Charging ground circuit voltage drop test
   D. Starter operating voltage test

5. Technician A says that a poor ground connection will cause reduced current flow in an electrical circuit. Technician B says that water corrosion in the wiring will cause reduced current flow in an electrical circuit. Who is correct?
   A. A only
   B. B only
   C. Both A and B
   D. Neither A nor B

6. A resistance test was performed on an open headlight switch and the result is 5.857 megohms. Technician A says that the switch is faulty because the reading is beyond the specifications. Technician B says that the switch has over 5 million ohms of resistance. Who is correct?

    A. A only

    B. B only

    C. Both A and B

    D. Neither A nor B

7. A truck is being diagnosed for a potential volt gauge problem. Technician A says that it is normal for the dash voltmeter to read a different voltage from that of a test voltmeter across the battery terminals. Technician B says that resistance in the dash voltmeter ground will affect its readings. Who is correct?

    A. A only

    B. B only

    C. Both A and B

    D. Neither A nor B

8. Technician A says that a short to ground before the load will cause the circuit protection device to open when the hot-side switch is turned on. Technician B says that a short to ground after the load will cause the circuit protection device to open when the hot-side switch is turned on. Who is correct?

    A. A only

    B. B only

    C. Both A and B

    D. Neither A nor B

9. Technician A says that a mistimed engine can cause slow cranking speed. Technician B says that low engine compression can cause rapid cranking speed. Who is correct?

    A. A only

    B. B only

    C. Both A and B

    D. Neither A nor B

10. Technician A says that any kind of wire can be used to replace a defective fusible link as long as the gauge is one size smaller than the circuit being protected. Technician B says that circuit breakers should always be replaced after repairing a short to ground problem. Who is correct?

    A. A only

    B. B only

    C. Both A and B

    D. Neither A nor B

11. A truck has a turn signal complaint. The left-front light does not work and the left-rear light flashes slower than normal. Technician A says the left-front bulb could be defective. Technician B says there could be an open circuit between the switch and the left-front bulb. Who is correct?

    A. A only

    B. B only

    C. Both A and B

    D. Neither A nor B

12. Which of the following would be the LEAST LIKELY test result of a good International Organization for Standardization (ISO) relay?

    A. 1 ohm when connected to terminals 30 and 85

    B. Out of limits (OL) when connected to terminals 30 and 87

    C. 80 ohms when connected to terminals 85 and 86

    D. 1 ohm when connected to terminals 30 and 87a

13. The temperature light stays on continuously while driving a heavy-duty truck. Technician A says that a shorted wire leading to the temperature sending unit could be the cause. Technician B says that an "open" temperature sending unit could be the cause. Who is correct?

    A. A only

    B. B only

    C. Both A and B

    D. Neither A nor B

14. A truck is being diagnosed for a "no communication" fault on the scan tool. Technician A says that an open terminating resistor could be the cause. Technician B says that a secure terminal connection at the splice block could be the cause. Who is correct?

    A. A only

    B. B only

    C. Both A and B

    D. Neither A nor B

15. A technician is preparing to load test the batteries on a truck with four batteries. Technician A says that trucks with multiple batteries should be tested with the batteries connected together. Technician B says that it is normal to see sparks while the load test is being performed. Who is correct?

    A. A only

    B. B only

    C. Both A and B

    D. Neither A nor B

16. Technician A says that measuring the terminal post voltage while the battery is being charged is a good way to determine the battery state of charge. Technician B says that terminal post voltage often drops down to approximately 7.5 volts while the engine is cranking. Who is correct?

    A. A only

    B. B only

    C. Both A and B

    D. Neither A nor B

17. The alternator needs to be replaced on a truck. Technician A says that the battery cables do not have to be disconnected while performing the exchange. Technician B says that the pulley diameter does not have to match when installing the new alternator. Who is correct?

    A. A only

    B. B only

    C. Both A and B

    D. Neither A nor B

18. Where is the low voltage disconnect (LVD) module connected to the truck?

    A. In series between the alternator and the battery pack
    B. In parallel with the battery pack and the truck power distribution center
    C. In parallel with the alternator and the battery pack
    D. In series between the battery pack and the truck power distribution center

19. A truck is being diagnosed for a slow crank problem. Technician A says that testing the voltage drop on the battery cables while cranking the engine will reveal cable problems. Technician B says that worn starter brushes could cause the slow crank problem. Who is correct?

    A. A only
    B. B only
    C. Both A and B
    D. Neither A nor B

20. A truck will not crank and the technician notices that the interior lights do not dim when the ignition switch is moved to the start position. Of the following, what is the most likely cause of this condition?

    A. Stuck closed starter relay
    B. Loose battery cable connections
    C. Starter mounting bolts loose
    D. Stuck open start switch

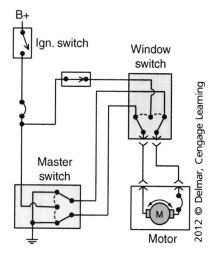

21. Referring to the figure above, the power window operates normally from the master switch, but does not operate using the window switch. Which of the following could be the cause?

    A. An open between the ignition switch and window switch
    B. An open in the window switch movable contacts
    C. An open in the master switch ground wire
    D. A short to ground at the circuit breaker in the motor

22. Which of the following would be the recommended method to use when professionally cleaning battery cable clamps?

    A. Pneumatic die grinder

    B. Wire brush

    C. Carburetor cleaner and scraper

    D. Screwdriver

23. All of the following parameters can be calibrated with a scan tool on some late-model trucks EXCEPT:

    A. Engine coolant temperature

    B. Maximum cruise control speed

    C. Maximum vehicle speed

    D. Maximum idle time

24. Technician A says that a broken wire to a door ajar switch will likely cause a parasitic draw problem. Technician B says that a blown horn fuse will likely cause a parasitic draw problem. Who is correct?

    A. A only

    B. B only

    C. Both A and B

    D. Neither A nor B

25. A truck is being diagnosed for an antilock brake system (ABS) indicator that is staying illuminated. All of the following steps are advisable to perform in the early stages of this diagnostic procedure EXCEPT:

    A. Use an ohmmeter to measure the resistance of all ABS speed sensors.

    B. Connect a scan tool to the truck to retrieve the diagnostic trouble code (DTC).

    C. Press the code activation switch to engage flash code diagnostics.

    D. Perform a visual inspection of the ABS system to identify obvious defects.

26. Technician A says that an electrical switch that has continuity will have reduced current flow. Technician B says that a piece of wire that has high resistance will have increased current flow. Who is correct?

    A. A only

    B. B only

    C. Both A and B

    D. Neither A nor B

27. Technician A says that a poor connection at the charging output wire can cause a charging problem. Technician B says that many charging circuits contain a circuit protection device. Who is correct?

    A. A only

    B. B only

    C. Both A and B

    D. Neither A nor B

28. The alternator belt is being tightened on a truck that uses V-belts. Technician A says that overtightening the alternator belt could cause bearing failure in the alternator. Technician B says that undertightening the alternator belt could cause bearing failure in the alternator. Who is correct?

    A. A only
    B. B only
    C. Both A and B
    D. Neither A nor B

29. A full-fielded alternator could cause all of the following conditions EXCEPT:

    A. High battery voltage level
    B. Battery gassing
    C. Low battery voltage level
    D. High alternator amperage output

30. A truck is being diagnosed for a charging problem. The alternator only charges at 11.8 volts. Technician A says that the charging output wire may have an excessive voltage drop. Technician B says that the charging ground circuit may have an excessive voltage drop. Who is correct?

    A. A only
    B. B only
    C. Both A and B
    D. Neither A nor B

31. Which of the following practices will reduce the chances of a battery cable connection getting corroded in the future?

    A. Using baking soda on the battery tray while the battery is removed
    B. Applying terminal protection spray after connecting the terminals to the batteries
    C. Applying bearing grease on the battery terminals prior to connecting the batteries
    D. Applying battery terminal cleaning spray after connecting the terminals to the batteries

32. A truck requires a new charging circuit connector at the alternator. Technician A says that the negative battery cable should be disconnected before beginning the repair. Technician B says that a butt loop connector is the recommended part for making the repair. Who is correct?

    A. A only
    B. B only
    C. Both A and B
    D. Neither A nor B

33. Which dual-filament bulb design has a round base with two contacts on the bottom and receives its ground through the case of the base?

    A. 1156
    B. 1157
    C. 3056
    D. 3057

34. Technician A says that the dash light dimmer is sometimes built into the headlight switch. Technician B says that the dimmer switch is sometimes built into the multi-function switch. Who is correct?

    A. A only

    B. B only

    C. Both A and B

    D. Neither A nor B

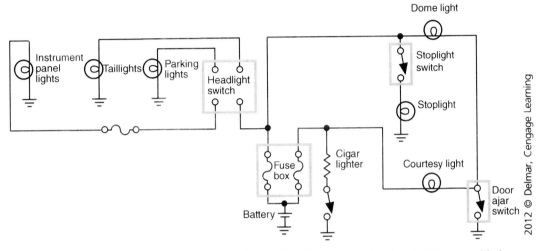

35. Referring to the figure above, the cigarette lighter fuse is blown in the circuit. The most likely result of this problem is:

    A. The courtesy and dome lights come on dimly when you push in the lighter.

    B. The stop and dome lights are completely inoperative.

    C. The parking lights, taillights, and instrument panel lights are inoperative.

    D. The dome light will not work when the door is opened.

36. A diode is being tested with a multimeter. Technician A says that a good diode test will read OL when the meter leads are connected in forward bias. Technician B says that a good diode will read approximately 0.5 volts when the meter leads are connected in reverse bias. Who is correct?

    A. A only

    B. B only

    C. Both A and B

    D. Neither A nor B

37. Technician A says that the hazard flasher on late-model vehicles can be replaced without disconnecting the vehicle battery. Technician B says that the turn signal flasher and hazard flasher are combined into one assembly on some late-model vehicles. Who is correct?

    A. A only

    B. B only

    C. Both A and B

    D. Neither A nor B

38. All of the following tools can be used to test a trailer light cord EXCEPT:

    A. Headlight with a jumper wire

    B. Ammeter

    C. 12 volt test light

    D. Trailer cord test tool

39. All of the following could cause an inaccurate gauge reading EXCEPT:

    A. A defective ground at the sender unit
    B. High battery voltage
    C. A defective instrument voltage regulator (IVR)
    D. High resistance in the gauge wiring

40. What type of sending unit does the fuel gauge use?

    A. Rheostat
    B. Transducer
    C. Photo resistor
    D. Thermistor

41. Which of the following details would most likely be located on a wiring diagram?

    A. Power and ground distribution for the circuit
    B. Location of the ground connection
    C. Updated factory information about pattern failures
    D. Flowchart for troubleshooting an electrical problem

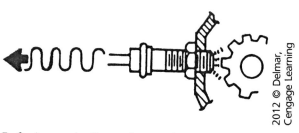

2012 © Delmar, Cengage Learning

42. Referring to the figure above, what can this type of sensor be used for?

    A. Speed sensing
    B. Pressure sensing
    C. Temperature sensing
    D. Level sensing

43. Which of the following wiper problems would LEAST LIKELY cause slow wiper operation?

    A. Faulty wiper speed relay
    B. Blown wiper fuse
    C. Poor alternator output
    D. Excessively tight wiper linkage

44. The left-side heated mirror does not clear the mirror as well as the right-side heated mirror. Which of the following conditions would most likely cause this problem?

    A. Faulty heated mirror switch
    B. Faulty right-side heated mirror
    C. Burned terminal at the driver's heated mirror electrical connector
    D. Weak alternator output

45. Technician A says that the starter solenoid contacts can be tested by performing a voltage drop test across the "bat" and "motor" terminals when the starter is disengaged. Technician B says that the pull-in winding can be tested by measuring the resistance from the "S" terminal to the "motor" terminal. Who is correct?

    A. A only
    B. B only
    C. Both A and B
    D. Neither A nor B

46. Which of the following conditions would LEAST LIKELY cause all of the power window motors to be inoperative?

    A. Faulty power window circuit breaker
    B. Bad ground at the master switch
    C. Binding power window regulator on the passenger side
    D. Blown power window fuse

47. Which of the following conditions takes place when the truck is plugged into a shore power supply plug?

    A. The truck is diagnosed for any engine electrical problems.
    B. The truck battery pack supplies power to the shore power system.
    C. The truck's electrical accessories receive power through the plug.
    D. The truck alternator supplies power to the shore power system.

48. Which of the following steps would most likely be performed when replacing the power door lock actuator?

    A. Remove the key lock tumbler.
    B. Remove the door panel.
    C. Remove the body control module (BCM).
    D. Remove the power window regulator.

49. While diagnosing a truck with electronic engine management and a "no-start" complaint, Technician A says that a digital multimeter could be used to check the electronic circuits. Technician B says that the DMM should be a high-impedance tool. Who is correct?

    A. A only
    B. B only
    C. Both A and B
    D. Neither A nor B

50. Which of the following devices would LEAST LIKELY signal the ECM to actuate the electric fan on a late-model heavy-duty truck?

    A. Trinary A/C switch
    B. Turbo boost sensor
    C. Coolant temperature sensor
    D. A/C pressure sensor

PREPARATION EXAM 5

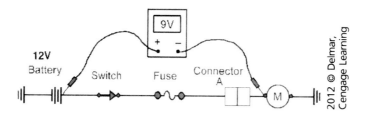

2012 © Delmar, Cengage Learning

1. Referring to the figure above, the voltmeter in the circuit reads 9 volts. Which of the following conditions would be the most likely cause of this measurement?

    A. Wire repair with a larger than specified wire
    B. Faulty bulb
    C. Bad bulb ground
    D. Burned terminal connection at connector A

2. All of the following conditions can cause reduced current flow in an electrical circuit EXCEPT:

    A. Loose terminal connections
    B. Hot-side wire rubbing a metal component
    C. Corrosion inside the wire insulation
    D. A wire that is too small

3. An electric fan on a late-model truck can be energized by which of the following methods?

    A. Viscous fluid switch
    B. Bi-metal spring switch
    C. Relay controlled by the engine control module (ECM)
    D. Vacuum thermal switch

4. Technician A says that a circuit with corrosion in the wiring will create a short circuit. Technician B says that a power wire that rubs a metal surface for a long period of time could create a short to ground. Who is correct?

    A. A only
    B. B only
    C. Both A and B
    D. Neither A nor B

5. All of the following problems could cause excessive key-off battery drain EXCEPT:

    A. Inverter switch stuck on
    B. Broken wire near the dome light socket
    C. Map light switch stuck closed
    D. Faulty rectifier bridge in the alternator

6. Which of the following devices would LEAST LIKELY be used as a circuit protection device?

    A. Maxi-fuse
    B. Circuit breaker
    C. Relay
    D. Positive temperature coefficient (PTC) thermistor

7. Technician A says that a loose drive belt could cause undercharging. Technician B says that undersized wiring between the alternator and the battery could cause undercharging. Who is correct?

    A. A only
    B. B only
    C. Both A and B
    D. Neither A nor B

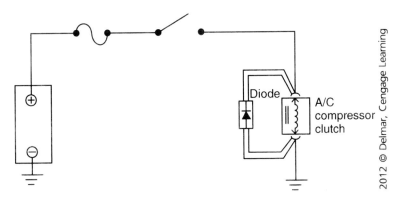

8. Referring to the figure above, Technician A says that the air conditioner (A/C) compressor clutch coil is directly connected to ground. Technician B says that the diode in the circuit protects the components from the voltage spike that is created as the coil is de-energized. Who is correct?

    A. A only
    B. B only
    C. Both A and B
    D. Neither A nor B

9. Technician A says that a blown ignition fuse can cause low cranking voltage. Technician B says that a weak battery can cause low cranking voltage. Who is correct?

    A. A only
    B. B only
    C. Both A and B
    D. Neither A nor B

10. A wiring schematic is being used to diagnose a truck electrical problem. Technician A says that most schematics show the wire colors of the wires. Technician B says that most schematics are drawn with the power coming from the top of the picture. Who is correct?

    A. A only
    B. B only
    C. Both A and B
    D. Neither A nor B

11. When diagnosing a repeated flasher failure, Technician A says that a 100 millivolt voltage drop problem may be the cause. Technician B says that proper grounding of the trailer sockets may correct the problem. Who is correct?

    A. A only

    B. B only

    C. Both A and B

    D. Neither A nor B

12. Technician A states that the scan tool receives data from a connector located on the ECM. Technician B states that the scan tool connects to the data bus using the data link connector (DLC). Who is correct?

    A. A only

    B. B only

    C. Both A and B

    D. Neither A nor B

13. The indicator for the bright lights does not work, but the bright and dim headlights function correctly. Which of the following conditions would be the most likely cause of this problem?

    A. Faulty headlight switch

    B. Faulty bulb socket for the bright indicator

    C. Faulty dimmer switch

    D. Faulty multi-function switch

14. A truck is in the repair shop for a battery problem. Technician A says that the battery voltage should be 12.4 volts before performing a battery load test. Technician B says that a battery can be accurately tested with a digital tester if it has at least 12 volts. Who is correct?

    A. A only

    B. B only

    C. Both A and B

    D. Neither A nor B

15. A truck is being diagnosed for a battery problem. Technician A says when disconnecting battery cables, always disconnect the negative cable first. Technician B says when connecting battery cables, always connect the negative cable first. Who is correct?

    A. A only

    B. B only

    C. Both A and B

    D. Neither A nor B

16. Technician A says that battery hold downs should always be installed to prevent batteries from bouncing, causing possible internal damage. Technician B states that a battery box need not be cleaned when replacing a battery because the case is insulated and the battery cannot discharge because of it. Who is correct?

    A. A only

    B. B only

    C. Both A and B

    D. Neither A nor B

17. A truck with a charging problem is being repaired. The technician finds that the charging output wire received damage and needs to be repaired. Technician A says to use weather-resistant connectors when making wire repairs in the engine compartment. Technician B says that solder and heat shrink is an acceptable method of wire repair in the engine compartment. Who is correct?

  A. A only
  B. B only
  C. Both A and B
  D. Neither A nor B

18. A starter circuit voltage drop test checks everything EXCEPT:

  A. Battery voltage
  B. Resistance in the positive battery cable
  C. Resistance in the negative battery cable
  D. Condition of the solenoid internal contacts

19. Which of the following functions would be LEAST LIKELY performed by the starter solenoid?

  A. Preventing the armature from overspinning
  B. Pushing the drive gear out to the flywheel
  C. Connecting the "bat" terminal to the "motor" terminal
  D. Providing a path for high current to flow

20. A starter is being removed from a truck. Technician A says that the negative battery cable should be removed prior to disconnecting the electrical connections at the starter. Technician B says that the fasteners should be removed prior to removing the electrical connections at the starter. Who is correct?

  A. A only
  B. B only
  C. Both A and B
  D. Neither A nor B

21. What is the function of the park circuit in the windshield wiper system?

  A. Return the wiper blades to the start position.
  B. Shut down the wiper motor in case of overheating.
  C. Help keep the wiper blades synchronized.
  D. Stop the wiper motor in case of a low-voltage problem.

22. A truck is being diagnosed for a slow crank. Smoke is noticed at the starter solenoid when the truck is in the crank mode. Technician A says that a loose attachment nut at the starter solenoid could be the cause. Technician B says that the connections at the starter solenoid should be checked for burned contacts. Who is correct?

  A. A only
  B. B only
  C. Both A and B
  D. Neither A nor B

23. A truck is being diagnosed for a starting problem. Technician A states that weak batteries can cause high current draw. Technician B states that faulty starter bushings can cause high current draw. Who is correct?

    A. A only

    B. B only

    C. Both A and B

    D. Neither A nor B

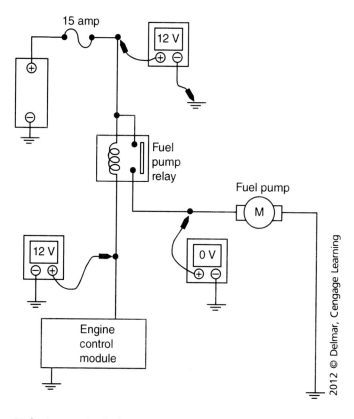

24. Referring to the fuel pump circuit in the figure above, the fuel pump is inoperative and the voltage readings were taken immediately after the circuit was closed. Technician A says that the fuel pump is likely defective. Technician B says that the relay coil could be open. Who is correct?

    A. A only

    B. B only

    C. Both A and B

    D. Neither A nor B

25. A truck's charging system produces 55 amps during a charging performance test. The charging specification for the truck is 110 amps. Technician A says that a full field test should be performed to see if the charging performance improves. Technician B says that the charging circuit voltage drop should be tested. Who is correct?

    A. A only

    B. B only

    C. Both A and B

    D. Neither A nor B

26. Which of the following devices would LEAST LIKELY be used as a spike suppression device for an electromagnetic coil?

    A. Resistor
    B. Diode
    C. Transistor
    D. Zener diode

27. An alternator output test is being performed. Technician A uses only a voltmeter connected across the battery positive terminal and negative terminal while the engine is running. Technician B says a carbon pile is not needed since the engine is already running. Who is correct?

    A. A only
    B. B only
    C. Both A and B
    D. Neither A nor B

28. A technician is testing alternator output. After starting the engine the test reveals that current output from the alternator slowly decreases the longer the engine runs. What can this mean?

    A. Alternator output is marginal; discontinue the test.
    B. The alternator drive belt is probably slipping on the pulley.
    C. The battery is slowly recovering to capacity.
    D. The diodes in the alternator are heating up and starting to fail.

29. A truck is being diagnosed for a charging problem. The alternator only charges at 12.4 volts. A voltage drop test is performed on the charging output wire and 1.8 volts are measured. Technician A says that a blown fusible link in the output circuit could be the cause. Technician B says that a loose nut at the charging output connector could be the cause. Who is correct?

    A. A only
    B. B only
    C. Both A and B
    D. Neither A nor B

30. Which of the following practices would be LEAST LIKELY followed when replacing the alternator on a heavy-duty truck?

    A. Use a box-end wrench to loosen the charging output nut.
    B. Remove the brushes from the old alternator and install them into the replacement alternator.
    C. Remove the drive belt prior to removing the alternator.
    D. Disconnect the negative battery cable prior to removing the alternator.

31. Technician A says that the low voltage disconnect (LVD) module is connected in series between the truck battery pack and the truck power distribution center. Technician B says that the LVD system will turn off battery power when the battery voltage level drops below 10.4 volts. Who is correct?

    A. A only
    B. B only
    C. Both A and B
    D. Neither A nor B

32. Technician A says that an overcharging alternator can cause lights that are brighter than normal. Technician B states that poor chassis grounds usually cause dim lights. Who is correct?

    A. A only
    B. B only
    C. Both A and B
    D. Neither A nor B

33. Technician A says that it is a good idea to coat the prongs and base of a new sealed-beam light assembly with dielectric grease before installing to prevent corrosion. Technician B says that white lithium grease can be used instead of dielectric grease. Who is correct?

    A. A only
    B. B only
    C. Both A and B
    D. Neither A nor B

34. The courtesy lights stay on continuously and have caused the batteries to be discharged while the truck is parked for long periods of time. Technician A says that the courtesy light switch could have a bad connection. Technician B says that a door ajar switch could be shorted. Who is correct?

    A. A only
    B. B only
    C. Both A and B
    D. Neither A nor B

35. A truck is being diagnosed for inoperative turn signals. All of the hazard lights work correctly. Technician A says that a blown turn signal bulb could be the cause. Technician B says that an open ground connection at the right-rear lamp socket could be the cause. Who is correct?

    A. A only
    B. B only
    C. Both A and B
    D. Neither A nor B

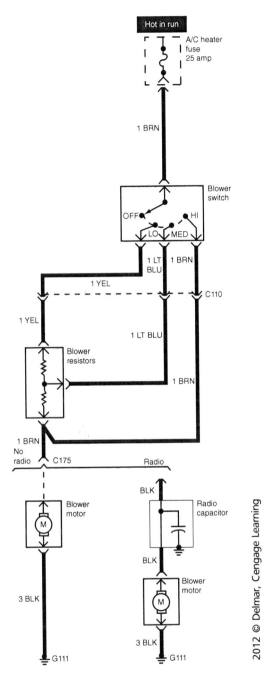

36. Referring to the figure above, which of the following statements best describes the wiring schematic?

  A. Power is sent to the light blue wire when the blower switch is set to medium speed.

  B. Power is sent to the yellow wire when the blower switch is set to the high speed.

  C. Power is sent to the brown wire when the blower switch is set to the low speed.

  D. The A/C heater fuse supplies ground to the blower motor circuit.

37. The back-up lights are inoperative on a truck. Technician A says that a faulty brake switch could be the cause. Technician B says that a stuck open back-up lamp switch could be the cause. Who is correct?

    A. A only
    B. B only
    C. Both A and B
    D. Neither A nor B

38. Technician A says that the main data link connector (DLC) is a square 9-pin connector. Technician B says that the scan tool should be used to engage flash code diagnostics on the truck. Who is correct?

    A. A only
    B. B only
    C. Both A and B
    D. Neither A nor B

39. All of the following types of electronic devices are used as inputs to electronic gauge assemblies EXCEPT:

    A. Thermistor
    B. Piezo resistor
    C. Diode
    D. Rheostat

40. On a truck with an electronic dash display, all of the gauge needles sweep from left to right immediately after turning the key on. Technician A says that this may indicate a fault with the instrument panel. Technician B states that this is due to high battery voltage. Who is correct?

    A. A only
    B. B only
    C. Both A and B
    D. Neither A nor B

41. All of the following methods can be used to retrieve codes from a truck's on-board computer EXCEPT:

    A. Laptop-based scan tool
    B. Self-contained scan tool
    C. Technical service bulletin
    D. Flash code diagnostics

42. The horn blows intermittently on a heavy-duty truck. Which of the following conditions would be the LEAST LIKELY cause of this problem?

    A. Wire rubbing a ground near the base of the steering column
    B. Sticking horn relay
    C. Faulty horn switch
    D. Horn fault

43. An electric horn on a medium-duty truck operates intermittently. Which of the following could be the cause?

    A. Blown fuse
    B. No power to relay
    C. Open in horn button circuit
    D. Defective horn relay

44. Which one of the following problems would be the most likely cause for inoperative windshield wipers?

    A. Stuck closed wiper switch
    B. Tripped thermal overload protector
    C. Closed circuit breaker
    D. High resistance in the motor wiring

45. Technician A says that the courtesy fuse should be removed prior to disconnecting the electrical connections at the starter. Technician B says that the starter can be supported by the connecting electrical wires without any expected damage to the wires. Who is correct?

    A. A only
    B. B only
    C. Both A and B
    D. Neither A nor B

46. Which of the following conditions would most likely cause the power windows to be inoperative?

    A. Missing ground connection to the master switch
    B. Faulty power lock switch
    C. Binding power window regulator
    D. Loose door panel

47. Which of the following conditions takes place when a truck is plugged into a shore power supply plug?

    A. The truck is diagnosed for any engine electrical problems.
    B. The battery pack receives an electrical charge.
    C. The truck battery pack supplies power to the shore power system plug.
    D. The truck alternator supplies power to the shore power system plug.

48. What kind of component is most likely used as a power lock actuator?

    A. Vacuum actuator
    B. Solenoid
    C. Electromagnetic motor
    D. Relay

49. Technician A says that power must be turned off in the circuit before using an ohmmeter to make a measurement. Technician B says that the ohmmeter applies a small amount of voltage to the circuit to calculate resistance. Who is correct?

A. A only

B. B only

C. Both A and B

D. Neither A nor B

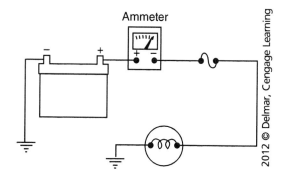

Ammeter

2012 © Delmar, Cengage Learning

50. Referring to the figure above, the ammeter indicates that current flow through the bulb is higher than specified. Which of the following could be the cause of this high current?

A. The fuse has an open circuit.

B. The battery voltage is low.

C. The light bulb filament is shorted.

D. The bulb filament has high resistance.

PREPARATION EXAM 6

1. A scan tool is connected to the engine computer of a late-model truck that has a check engine light staying on. Technician A says that the system should be checked for diagnostic trouble codes (DTCs). Technician B says that the data list should be checked for irregular readings. Who is correct?

A. A only

B. B only

C. Both A and B

D. Neither A nor B

2. How is an ammeter connected in order to read live current?

A. In parallel with the circuit with the power turned off

B. In parallel with the circuit with the power turned on

C. In series with the circuit with the power turned off

D. In series with the circuit with the power turned on

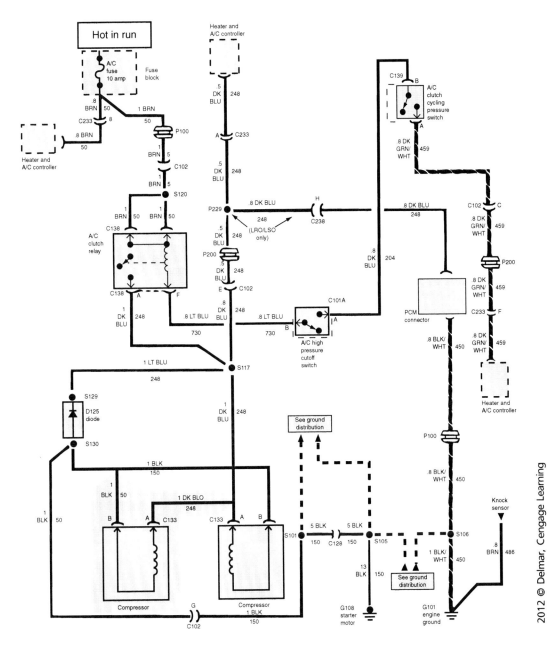

3. Referring to the figure above, this electrical schematic is being examined by two technicians. Technician A says that the schematic shows the heater and air conditioner (A/C) controller circuit. Technician B says that this circuit operates with circuit #50 (1 brn) being open. Who is correct?

   A. A only

   B. B only

   C. Both A and B

   D. Neither A nor B

4. Technician A says that a stuck closed dome light switch could cause excessive key-off drain. Technician B says that a faulty starter solenoid could cause excessive key-off drain. Who is correct?

   A. A only

   B. B only

   C. Both A and B

   D. Neither A nor B

5. A truck with a key-off battery drain problem is being diagnosed. Technician A states that having zero key-off battery drain is common on a truck equipped with multiple control modules. Technician B says battery drain can be caused by excess moisture on top of the battery. Who is correct?

   A. A only

   B. B only

   C. Both A and B

   D. Neither A nor B

6. A fusible link needs to be replaced. Technician A says that the batteries should be disconnected first. Technician B says that the same gauge fuse link wire should be used in the repair. Who is correct?

   A. A only

   B. B only

   C. Both A and B

   D. Neither A nor B

7. An alternator is overcharging. Technician A says that this can only be caused by a defective voltage regulator. Technician B states that excessive resistance in the charging circuit wiring can cause this overcharging. Who is correct?

   A. A only

   B. B only

   C. Both A and B

   D. Neither A nor B

8. Technician A says that resistors are sometimes used as spike suppression devices in relays. Technician B says that diodes are sometimes used as spike suppression devices in relays. Who is correct?

   A. A only

   B. B only

   C. Both A and B

   D. Neither A nor B

9. Which of the following conditions would LEAST LIKELY cause a slow crank problem?

   A. Mistimed engine

   B. Worn starter brushes

   C. Battery with a higher CCA rating than original specifications

   D. Engine with tight main bearings

10. Technician A says that a solenoid can stick in the applied position if the return spring breaks. Technician B says that a solenoid has a coil that can be tested with an ohmmeter for the correct resistance. Who is correct?

    A. A only
    B. B only
    C. Both A and B
    D. Neither A nor B

11. The stoplights do not work when the brake pedal is depressed on a heavy-duty truck. Which of the following conditions would be the most likely cause of this problem?

    A. Faulty cruise control switch
    B. Damaged air hose at the stoplight switch
    C. Misadjusted air governor
    D. Faulty hazard flasher

12. Technician A says that a wiring diagram uses schematic symbols to represent electrical components. Technician B says that wiring diagrams usually show the splice and connector numbers for each circuit. Who is correct?

    A. A only
    B. B only
    C. Both A and B
    D. Neither A nor B

13. What is the function of a pyrometer?

    A. Monitors fuel temperature
    B. Indicates the battery state of charge
    C. Indicates engine speed
    D. Monitors exhaust temperature

14. All of the following statements about the J1939 data bus network are correct EXCEPT:

    A. The J1939 data bus is a two-wire network.
    B. The J1939 data bus can be tested at the 9-pin data connector.
    C. The J1939 data bus should have 60 ohms of resistance when tested with an ohmmeter.
    D. The J1939 data bus can communicate at 250 gigabytes per second.

15. When checking open circuit battery voltage, Technician A says that a 12 volt battery is considered fully charged if a voltmeter probed across it reads anything over 12 volts. Technician B states that the charging system should produce approximately 13.5 to 14.5 volts. Who is correct?

    A. A only
    B. B only
    C. Both A and B
    D. Neither A nor B

16. Technician A says that battery boxes should be cleaned with mineral spirits. Technician B says that battery hold down hardware should be tight and secure to prevent the battery from moving during use. Who is correct?

    A. A only

    B. B only

    C. Both A and B

    D. Neither A nor B

17. The alternator needs to be replaced on a heavy-duty truck. Technician A says that the negative battery cable should be removed during this process. Technician B says the drive belt should be closely inspected during this process. Who is correct?

    A. A only

    B. B only

    C. Both A and B

    D. Neither A nor B

18. Which would be the last connection to make when connecting jumper cables to a truck with dead batteries?

    A. Positive post of the truck with the dead battery

    B. Positive post of the truck with the charged battery

    C. Negative post of the truck with the charged battery

    D. Engine block of the truck with the dead battery

19. How is a starter ground circuit resistance check performed?

    A. A voltmeter is connected between the ground terminal of the battery and starter ground stud and read while the engine is being cranked.

    B. An ohmmeter is connected between the starter relay housing and the starter housing.

    C. An ohmmeter is connected between the ground side of the battery and the starter housing and read while the engine is cranking.

    D. A voltmeter is connected between the positive side of the battery and the starter solenoid while the engine is off.

20. Which of the following components is LEAST LIKELY to be part of the starter control circuit?

    A. Positive battery cable

    B. Ignition switch

    C. Park/neutral switch

    D. Starter relay

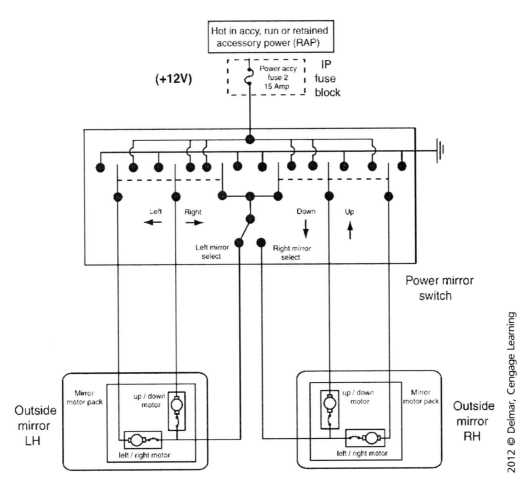

21. Referring to the figure above, the left-side mirror works correctly but the right-side mirror is totally inoperative. Which of the following conditions would most likely be the cause of this problem?

   A. Faulty right-side up/down motor

   B. Broken contact at the mirror select switch

   C. Missing ground at the power mirror switch

   D. Blown power accessory fuse

22. Technician A says that the replacement starter assembly should be bench tested prior to installing it on the engine. Technician B says that the starter connectors and terminals should be inspected and cleaned prior to installing a replacement starter. Who is correct?

   A. A only

   B. B only

   C. Both A and B

   D. Neither A nor B

23. The battery housing received some damage from driving a vehicle on rough roads. Electrolyte spilled all over the battery tray. Technician A says that carburetor cleaner should be used to clean the area. Technician B says that baking soda could be used to neutralize the battery acid. Who is correct?

   A. A only

   B. B only

   C. Both A and B

   D. Neither A nor B

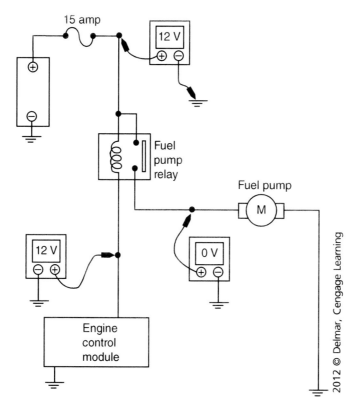

24. Referring to the figure above, the fuel pump is inoperative and the voltage readings shown were taken immediately after the circuit was closed. Which of the conditions listed below is the most likely cause?

   A. Open 15 amp fuse

   B. Defective engine control module (ECM)

   C. Defective fuel pump relay

   D. Defective fuel pump

25. Technician A says that the battery voltage can drop to 6 volts while cranking during normal conditions. Technician B says that one weak battery can cause a battery bank to perform poorly during extremely cold weather. Who is correct?

   A. A only

   B. B only

   C. Both A and B

   D. Neither A nor B

26. All of the following devices could be used as a circuit protection device EXCEPT:

    A. Maxi-fuse
    B. Circuit breaker
    C. Positive temperature coefficient (PTC) thermistor
    D. Jumper wire

27. A vehicle is being diagnosed for a charging problem. The alternator produced 125 amps during the output test and is rated at 130 amps. Technician A says that the alternator could have a bad diode. Technician B says that the alternator drive pulley could be too large in diameter. Who is correct?

    A. A only
    B. B only
    C. Both A and B
    D. Neither A nor B

28. Technician A says that when performing an alternator output test, a voltmeter should be connected in series with the alternator output terminal and the battery ground cable. Technician B says that a carbon pile should be used when performing an alternator output test. Who is correct?

    A. A only
    B. B only
    C. Both A and B
    D. Neither A nor B

29. The ground side of a truck charging circuit is being tested for voltage drop. Technician A says to place the voltmeter leads on the voltage regulator ground terminal and the vehicle battery. Technician B says to place the voltmeter leads on the alternator negative terminal and the battery ground terminal. Who is correct?

    A. A only
    B. B only
    C. Both A and B
    D. Neither A nor B

30. The charging system is being inspected during a maintenance inspection. The technician performs a voltage drop test on the charging system and finds 0.2 volts on the positive wire and finds 0.02 volts on the ground wire. Technician A says that a defective voltage regulator could cause this condition. Technician B says that defective alternator brushes could cause this condition. Who is correct?

    A. A only
    B. B only
    C. Both A and B
    D. Neither A nor B

31. Technician A says that batteries can be recharged more quickly by using a low setting on the battery charger. Technician B says that batteries can be more thoroughly charged by using a high setting on the battery charger. Who is correct?

    A. A only
    B. B only
    C. Both A and B
    D. Neither A nor B

32. A truck is being diagnosed for a charging problem. Technician A says that most charging circuits include a circuit protection device. Technician B says that all repairs should be resistant to water intrusion. Who is correct?

    A. A only
    B. B only
    C. Both A and B
    D. Neither A nor B

33. A headlight aiming procedure is being performed. Technician A says that when headlight aiming equipment is not available, headlight aiming can be checked by projecting the high beam of each light upon a screen or chart at a distance of 25 feet ahead of the headlights. Technician B says that when aiming the headlights, the vehicle should be exactly parallel to the chart or screen. Who is correct?

    A. A only
    B. B only
    C. Both A and B
    D. Neither A nor B

34. The headlights work on high beams but are inoperative on low beams. Technician A says that the dimmer switch could be the fault. Technician B says that both bulbs could be faulty. Who is correct?

    A. A only
    B. B only
    C. Both A and B
    D. Neither A nor B

35. When the driver door of the vehicle is opened, the interior lights illuminate, but very dimly. Technician A says that the headlight switch may be defective. Technician B says a grounding problem to the door switch may be the problem. Who is correct?

    A. A only
    B. B only
    C. Both A and B
    D. Neither A nor B

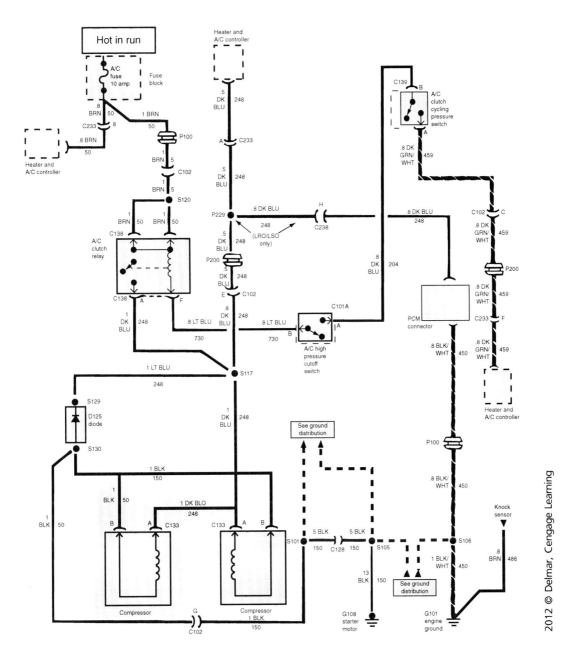

36. Referring to wiring schematic in the figure above, all of the following statements are correct EXCEPT:

   A. Circuit #450 uses a black wire with a white tracer.

   B. Circuit #150 uses a black wire.

   C. Circuit #248 uses a white wire.

   D. Circuit #50 uses a brown wire.

37. Turn signals are being diagnosed. Technician A says that a faulty hazard flasher assembly could cause the turn signals to be inoperative. Technician B says that hazard lights and turn signals use different bulbs and sockets. Who is correct?

    A. A only
    B. B only
    C. Both A and B
    D. Neither A nor B

38. All of the following statements are correct concerning turn signal operation on a heavy-duty truck EXCEPT:

    A. Some turn signal bulbs share a filament with the stoplights.
    B. Some turn signal bulbs use the same socket as the back-up lights.
    C. The turn signal flasher is sometimes combined with the hazard flasher.
    D. Some turn signal systems incorporate a relay to make the turn signals flash correctly.

39. All the gauges are erratic in an instrument panel with thermal-electric gauges and an instrument voltage limiter. Technician A says the alternator may be at fault. Technician B says the instrument voltage limiter may be defective. Who is correct?

    A. A only
    B. B only
    C. Both A and B
    D. Neither A nor B

40. All of the following conditions could cause the temperature gauge to read "high" EXCEPT:

    A. Shorted temperature sending unit
    B. Grounded temperature sending unit wire
    C. Open instrument panel fuse
    D. Faulty instrument panel circuit board

41. Technician A says that a high-impedance digital meter is needed to perform voltage tests on the data bus network. Technician B says that an oscilloscope can be used to view the communication activity on the data bus network. Who is correct?

    A. A only
    B. B only
    C. Both A and B
    D. Neither A nor B

42. Which of the following tools could be used to test a coolant temperature sending unit?

    A. Terminal removal tool
    B. Jumper wire
    C. Digital multimeter (DMM)
    D. Test light

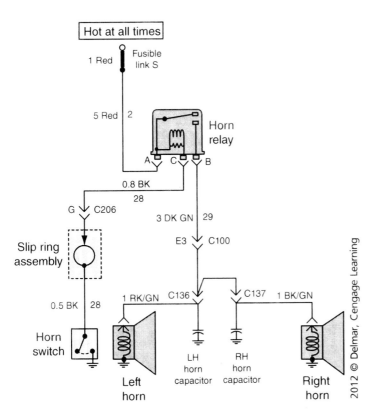

43. Referring to the figure above, the problem is a horn that sounds continuously. Technician A says that a short to ground at connector C206 may cause this. Technician B states that a short to ground at connector C100 could cause this problem. Who is correct?

    A. A only

    B. B only

    C. Both A and B

    D. Neither A nor B

44. A wiper motor fails to operate. Which of the following would be the LEAST LIKELY cause?

    A. Defective wiper switch

    B. Tripped thermal overload protector

    C. Tripped circuit breaker

    D. High resistance in the motor wiring

45. The starter solenoid performs all of the following functions EXCEPT:

    A. Provides a path for high current to flow into the starter

    B. Pushes the drive gear out to the flywheel

    C. Connects the "bat" terminal to the "motor" terminal

    D. Provides gear reduction to increase torque in the starter

46. What is the most likely cause for a windshield wiper system that only works on low speed?

    A. Blown fuse

    B. Faulty multi-function switch

    C. Loose ground at the wiper motor

    D. Open park switch

47. All of the following wiper problems could cause slow wiper operation EXCEPT:

    A. Weak batteries

    B. Poor alternator output

    C. Worn wiper blades

    D. Excessively tight wiper linkage

48. A truck has a windshield washer pump system that does not operate properly. Which of the following conditions would be the most likely cause of this problem?

    A. Binding wiper linkage

    B. Faulty wiper park switch

    C. Faulty windshield wiper motor

    D. Clogged washer hose

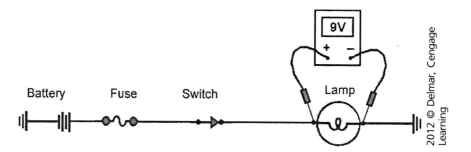

49. Referring to the figure above, the 12 volt battery is fully charged and the switch is closed. Which of the following conditions would be the most likely cause of this measurement?

    A. Wire repair that used a larger than specified wire

    B. Faulty bulb

    C. Blown fuse

    D. Loose terminal connection at the switch

50. The auxiliary power outlet is inoperative and the fuse is found to be open. What is the LEAST LIKELY cause for this condition?

    A. Shorted power wire near the auxiliary connector

    B. Foreign metal object in the auxiliary power outlet

    C. Faulty electrical device connected to the outlet

    D. Open internal connection at the power outlet

Answer Keys and Explanations

INTRODUCTION

Included in this section are the answer keys for each preparation exam, followed by individual, detailed answer explanations and a reference identifying the designated task area being assessed by each specific question. This additional reference information may prove useful if you need to refer back to the task list located in Section 4 of this book for additional support.

PREPARATION EXAM 1—ANSWER KEY

1.	B	18.	C	35.	C
2.	B	19.	D	36.	B
3.	A	20.	A	37.	C
4.	B	21.	A	38.	A
5.	D	22.	D	39.	D
6.	B	23.	B	40.	C
7.	A	24.	B	41.	B
8.	C	25.	D	42.	C
9.	D	26.	C	43.	B
10.	C	27.	A	44.	C
11.	A	28.	B	45.	B
12.	C	29.	A	46.	B
13.	C	30.	C	47.	B
14.	C	31.	C	48.	A
15.	C	32.	D	49.	D
16.	D	33.	D	50.	C
17.	B	34.	C		

PREPARATION EXAM 1—EXPLANATIONS

TASK A.3

1. A resistance test was performed on an open toggle switch and the result is 5.857 megohms. Technician A says that the switch is faulty because the reading is beyond the specifications. Technician B says that the switch has over 5 million ohms of resistance. Who is correct?

    A.   A only
    B.   B only
    C.   Both A and B
    D.   Neither A nor B

    Answer A is incorrect. An open switch is supposed to have very high resistance.

    Answer B is correct. Only Technician B is correct. The reading is over 5 million ohms. An open switch should have very high resistance.

    Answer C is incorrect. Only Technician B is correct.

    Answer D is incorrect. Technician B is correct.

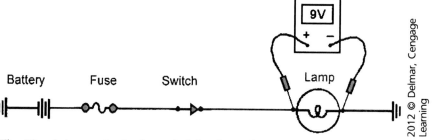

Battery   Fuse   Switch   Lamp

2012 © Delmar, Cengage Learning

TASK A.1

2. The 12 volt battery in the figure is fully charged and the switch is closed. Which of the following conditions would be the most likely cause of this measurement?

    A.   Larger than specified wire used in wire repair
    B.   Loose ground for the bulb
    C.   Blown fuse
    D.   Faulty bulb

    Answer A is incorrect. A larger wire used in a wire repair would not cause a voltage loss.

    Answer B is correct. A loose ground for the bulb could cause a voltage loss at the ground connection. Performing a voltage drop test on the ground circuit could isolate this problem.

    Answer C is incorrect. A blown fuse would prevent any voltage from getting past the fuse contacts.

    Answer D is incorrect. There is no evidence that the bulb is faulty. The meter shows only 9 volts being dropped across the bulb.

TASK E.6

3. Which of the following conditions could cause the power locks to fail to operate?

    A.   Broken connector at the power lock switch
    B.   Closed power lock circuit breaker
    C.   An oversized fuse
    D.   An oversized wire repair in the power lock circuit

    Answer A is correct. A broken connector at the power lock switch could cause the power locks to be inoperative because it would likely create an open circuit.

    Answer B is incorrect. A closed power lock circuit breaker would allow normal operation of the power door locks.

    Answer C is incorrect. An oversized fuse would allow the power door locks to function. For safety reasons, however, a technician should never install a fuse that has a larger rating than the specification.

    Answer D is incorrect. An oversized wire would not hinder the operation of the power door locks.

4. When using a 12 volt, unpowered test lamp to test for voltage, all of the following are true EXCEPT:

    A. The test lamp may be connected to the battery ground.

    B. The battery should first be disconnected.

    C. The test lamp may be connected to a chassis ground.

    D. The test lamp will not illuminate when probed into the circuit at a point after the load.

TASK A.1

Answer A is incorrect. A 12 volt unpowered test lamp can be connected to the battery ground cable or any quality ground on the truck.

Answer B is correct. The test light would not operate if the battery were disconnected.

Answer C is incorrect. A 12 volt unpowered test lamp can be connected to the chassis ground or any quality ground on the truck.

Answer D is incorrect. The voltage level will drop to near zero at points past the load, causing the test lamp to not illuminate if probed into this part of the circuit.

5. After removing a 30 amp circuit breaker from a fuse panel, a technician checks continuity across its terminals with a digital multimeter (DMM). Technician A says that the current from the ohmmeter will open the circuit breaker. Technician B says the ohmmeter should read infinity if the circuit breaker is good. Who is correct?

TASK A.3

    A. A only

    B. B only

    C. Both A and B

    D. Neither A nor B

Answer A is incorrect. The small amount of current supplied by an ohmmeter would not cause a good circuit breaker to open up.

Answer B is incorrect. The ohmmeter should read low resistance on a good circuit breaker.

Answer C is incorrect. Neither Technician is correct.

Answer D is correct. Neither Technician is correct. Using an ohmmeter to test a circuit breaker will not cause the contacts to open up because the current from the meter will be very small. A good circuit breaker will have a low reading when being tested with an ohmmeter.

6. Technician A uses a test lamp to detect resistance. Technician B uses a jumper wire when verifying the operation of circuit breakers, relays, and switches. Who is correct?

TASK A.3

    A. A only

    B. B only

    C. Both A and B

    D. Neither A nor B

Answer A is incorrect. A test lamp would not be a useful tool when trying to test for resistance. The test lamp can be used on some electrical circuits as a method to test for voltage being present in a circuit. The test lamp is not very accurate. It should not be used on data circuits because computer damage could result from excessive current flow.

Answer B is correct. Only Technician B is correct. A jumper wire can sometimes be used when testing electrical components, such as circuit breakers, relays, and switches.

Answer C is incorrect. Only Technician B is correct.

Answer D is incorrect. Technician B is correct.

TASK C.1

7. Technician A says that a charge indicator light should come on with the key on, engine off. Technician B states that if the indicator is not on with the engine running, it can be assumed that the charging system is functioning properly. Who is correct?

    A. A only
    B. B only
    C. Both A and B
    D. Neither A nor B

Answer A is correct. Technician A is correct. Most systems operate by turning on the charge warning light with the key on, engine off.

Answer B is incorrect. Even when the indicator light is off with the engine running, the alternator may still not be able to handle all of the electrical load if alternator output is not up to specification.

Answer C is incorrect. Only Technician A is correct.

Answer D is incorrect. Technician A is correct.

TASK A.5

8. Technician A says that an amp clamp is a useful tool to use when checking for key-off drain. Technician B says that a faulty rectifier bridge in the alternator could cause excessive key-off drain. Who is correct?

    A. A only
    B. B only
    C. Both A and B
    D. Neither A nor B

Answer A is incorrect. Technician B is also correct.

Answer B is incorrect. Technician A is also correct.

Answer C is correct. Both Technicians are correct. Amp clamps allow a technician to install the tool around the positive or negative battery cables and accurately measure current flow. The advantage to using this tool is that the battery does not have to be disconnected to install the ammeter. If a diode shorts out in the rectifier bridge in the alternator, then electrical current can flow from the battery into the alternator and cause the battery to run down.

Answer D is incorrect. Both Technicians are correct.

TASK B.15

9. All of these conditions would cause the starter not to crank the engine EXCEPT:

    A. The battery is not connected to the starter motor.
    B. The solenoid does not engage the starter drive pinion with the engine flywheel.
    C. The control circuit fails to switch the large-current circuit.
    D. The starter drive pinion fails to disengage from the flywheel.

Answer A is incorrect. If the battery cable were not connected to the starter motor, the starter would not rotate.

Answer B is incorrect. If the solenoid does not engage the pinion with the engine flywheel, the engine will not crank.

Answer C is incorrect. The control circuit is supposed to switch the large-current circuit in order to crank the engine. This is the function of the starter solenoid and the magnetic switch.

Answer D is correct. A drive pinion that fails to disengage will not cause an engine to not start. When the engine starts, the flywheel spins the pinion faster than the armature. This action releases the rollers, unlocking the pinion gear from the armature shaft. The pinion then "overruns" the armature shaft freely until being pulled out of the mesh without stressing the starter motor. Note that the overrunning clutch is moved in and out of mesh with the flywheel by linkage operated by the solenoid.

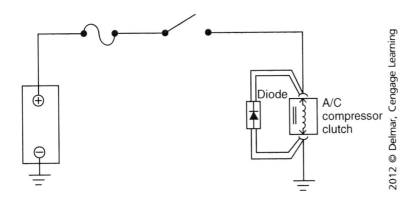

10. Referring to the figure above, what is the diode being used for?

    A. Allowing the A/C compressor to run in one direction only
    B. Protecting the A/C compressor clutch from a voltage spike
    C. Protecting the circuit from a possible voltage spike
    D. Limiting the current flow to the A/C compressor clutch

    TASK A.7

    Answer A is incorrect. The compressor clutch can only turn in one direction, regardless of the application of the clutch. The driving engine cannot turn backward.

    Answer B is incorrect. It is the A/C circuit that is being protected, not the clutch itself. The voltage spike is redirected back into the clutch during disengagement.

    Answer C is correct. The purpose of the diode is to protect the rest of the circuit from a voltage spike generated when the clutch magnetic field collapses.

    Answer D is incorrect. The diode cannot limit current flow to the clutch because a diode is simply a one-way check valve, not a resistor. Even if the diode acted as a resistor, it would have to be wired in series with the clutch, not in parallel, to affect the current flow.

11. The stoplights are inoperative on a truck equipped with air brakes. Technician A says that a faulty pressure switch could be the cause. Technician B says that a faulty hazard lamp switch could be the cause. Who is correct?

    A. A only
    B. B only
    C. Both A and B
    D. Neither A nor B

    TASK D.7

    Answer A is correct. Only Technician A is correct. The stoplights on an air-brake-equipped truck are activated by a pressure switch in the apply circuit of the air-brake system.

    Answer B is incorrect. A faulty hazard lamp switch could not cause the stoplights to be inoperative.

    Answer C is incorrect. Only Technician A is correct.

    Answer D is incorrect. Technician A is correct.

TASK A.8

12. Technician A says that a solenoid has a movable metallic plunger that creates linear movement when the coil is energized. Technician B says that some solenoids are used to control fluid in hydraulic circuits. Who is correct?

    A. A only
    B. B only
    C. Both A and B
    D. Neither A nor B

    Answer A is incorrect. Technician B is also correct.

    Answer B is incorrect. Technician A is also correct.

    Answer C is correct. Both Technicians are correct. Solenoids have a movable plunger that creates linear movement. This movement can be used to perform work. Examples of solenoids like this include the starter solenoid and the power lock solenoid. In addition, solenoids can be used to control hydraulic circuits. Examples of these solenoids include fuel injectors, pressure control solenoids, and transmission shift solenoids.

    Answer D is incorrect. Both Technicians are correct.

TASK E.4

13. The charging light stays on while driving. Technician A says that a grounded wire near the alternator could be the cause. Technician B says that a faulty circuit in the instrument cluster could be the cause. Who is correct?

    A. A only
    B. B only
    C. Both A and B
    D. Neither A nor B

    Answer A is incorrect. Technician B is also correct.

    Answer B is incorrect. Technician A is also correct.

    Answer C is correct. Both Technicians are correct. A grounded wire near the alternator could cause the charge light to stay on. A faulty circuit in the instrument cluster could also cause the charge light to stay on.

    Answer D is incorrect. Both Technicians are correct.

TASK A.3

14. All of the following electrical tools could be used to diagnose a data bus network problem EXCEPT:

    A. Oscilloscope
    B. Digital multimeter
    C. Analog voltmeter
    D. Scan tool

    Answer A is incorrect. An oscilloscope could be used to view the electrical signals being transmitted on the data bus network.

    Answer B is incorrect. A DMM could be used to measure the voltage levels on the data bus network.

    Answer C is correct. An analog voltmeter should not be used because most analog meters do not have high impedance.

    Answer D is incorrect. A scan tool could be used to communicate with the computers that use the data bus network.

15. How many load test amps are pulled from the battery during a high rate discharge (load) test?

    A. Two times the cold cranking ampere (CCA) task rating of the battery

    B. One and one-half times the CCA rating of the battery

    C. One-half the CCA rating of the battery

    D. Three-fourths the CCA rating of the battery

TASK B.2

Answer A is incorrect. Two times the cold cranking ampere (CCA) rating of the battery would be too much load to put on a battery during a load test.

Answer B is incorrect. One and one-half times the CCA rating would be too much load to put on a battery during a load test.

Answer C is correct. One-half the CCA rating load should be applied to a battery for 15 seconds to run a correct load test.

Answer D is incorrect. Three-fourths of the CCA rating load would be too much load to put on a battery during a load test.

16. A customer has a truck towed to the shop and says that the starter would not crank the engine. What should be checked first?

    A. Ground cable connection

    B. Starter solenoid circuit

    C. Ignition switch crank circuit

    D. Battery for proper charge

TASK B.8

Answer A is incorrect. The ground cable connection is important to the operation of the truck electrical system, but it would not typically be the first step of diagnosing a no crank problem.

Answer B is incorrect. The starter solenoid circuit may have to be tested at some point during a no crank diagnosis, but it would not typically be checked first.

Answer C is incorrect. The ignition switch crank circuit may have to be tested at some point during a no crank diagnosis, but it would not typically be checked first.

Answer D is correct. The battery should be inspected as one of the first steps during a no crank diagnosis.

17. An alternator is being replaced on a heavy-duty truck. Technician A uses an air tool to remove the fastening nut from the charging output wire. Technician B disconnects the negative battery cable prior to removing the alternator. Who is correct?

    A. A only

    B. B only

    C. Both A and B

    D. Neither A nor B

TASK C.6

Answer A is incorrect. An air tool should never be used to remove the fastening nut from the charging output wire. Doing this could damage the internal connections of the alternator.

Answer B is correct. Only Technician B is correct. It is a good practice to remove the negative battery cable prior to removing the alternator.

Answer C is incorrect. Only Technician B is correct.

Answer D is incorrect. Technician B is correct.

TASK B.5

18. Technician A says that batteries can be recharged more quickly by using a high setting on the battery charger. Technician B says that batteries can be charged more thoroughly by using a low setting on the battery charger. Who is correct?

   A. A only

   B. B only

   C. Both A and B

   D. Neither A nor B

Answer A is incorrect. Technician B is also correct.

Answer B is incorrect. Technician A is also correct.

Answer C is correct. Both Technicians are correct. Batteries can be charged on a high setting more quickly. The technician should monitor the voltage and temperature of the battery in order to not overcharge the battery, however. A slow charge typically provides a more thorough charge, but it takes more time to do.

Answer D is incorrect. Both Technicians are correct.

TASK B.9

19. A heavy-duty truck is slow to crank. Technician A says that testing the voltage drop on the positive battery cable while cranking the engine could reveal potential problems in the battery. Technician B says that testing the voltage drop on the negative battery cable must be done with the ignition switch in the off position. Who is correct?

   A. A only

   B. B only

   C. Both A and B

   D. Neither A nor B

Answer A is incorrect. Performing a voltage drop test on the battery cables will test the battery cable, not the battery.

Answer B is incorrect. Voltage drop tests on any wire or cable need to be done while the circuit is activated.

Answer C is incorrect. Neither Technician is correct.

Answer D is correct. Neither Technician is correct. Testing the voltage drop on the battery cables is done while cranking the engine; the specification is usually 0.5 volts or less.

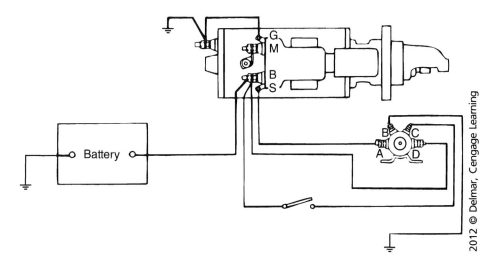

20. Referring to the figure above, which component is being checked if terminals C and D are jumped at the magnetic switch?

    A. Starting switch
    B. Battery
    C. Magnetic switch
    D. Starter

TASK B.10

Answer A is correct. The starting switch is being by-passed in this test, so its operation is being tested.

Answer B is incorrect. Although the battery is being tested here, it is not the objective of this test.

Answer C is incorrect. When terminals C and D are jumped, the magnetic switch is activated. This test by-passes the start switch and should cause the starter to engage.

Answer D is incorrect. Although the starter is activated, it is not the objective of this test.

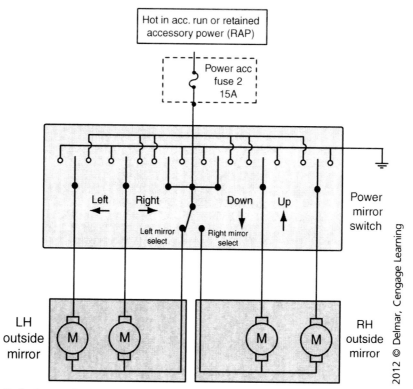

TASK E.11

21. Referring to the figure above, the left mirror functions normally, but the right-side power mirror does not function at all. Technician A says that mirror select switch could be defective. Technician B says that the power mirror switch could have a disconnected ground. Who is correct?

A. A only
B. B only
C. Both A and B
D. Neither A nor B

Answer A is correct. Only Technician A is correct. A defective mirror select switch could cause the right-side mirror to not operate at all.

Answer B is incorrect. A disconnected ground on the power mirror switch would cause all power mirror operation to cease.

Answer C is incorrect. Only Technician A is correct.

Answer D is incorrect. Technician A is correct.

TASK B.14

22. Which of the following steps should the technician do first when beginning to remove a starter from a truck?

A. Pull the starter fuse at the power distribution center.
B. Remove the starter bolts.
C. Disconnect the wires at the starter.
D. Remove the negative battery cable and tape the terminal.

Answer A is incorrect. It is typically not necessary to remove any fuses while removing a starter from a vehicle.

Answer B is incorrect. Removing the starter bolts before disconnecting the battery is not advised due to safety issues. The battery should always be disconnected as the first step.

Answer C is incorrect. Removing the wires at the starter before disconnecting the battery is a dangerous practice because the positive battery cable connects directly to the starter. A short to ground could easily occur if the battery is not disconnected first.

Answer D is correct. The negative battery cable should always be removed prior to disconnecting and removing a starter.

23. A technician is working on a truck and finds that a battery cable terminal end is badly corroded. All of the following are proper repair procedures EXCEPT:

    A. Replace the entire cable assembly.

    B. Replace the terminal with a bolt-on end and heat-shrink tubing.

    C. Replace the terminal with a crimp-on end and heat-shrink tubing.

    D. Replace the terminal with a soldered end and heat-shrink tubing.

TASK B.13

    Answer A is incorrect. Replacing the entire battery cable is an acceptable method of repair. It may be the fastest and most economical way, depending on shop preferences.

    Answer B is correct. You should never replace a battery cable end with an aftermarket type bolt-on end. They have poor wire contact characteristics and the entire joint is exposed and subject to corrosion.

    Answer C is incorrect. Replacing a battery cable terminal end with a crimp-on terminal and heat-shrink tubing is considered an acceptable method of repair.

    Answer D is incorrect. Replacing a battery cable terminal end with a soldered-on terminal and heat-shrink tubing is considered an acceptable method of repair.

24. While checking the fuses in a tractor/trailer, the technician finds an open fuse. Which of the following is the next step?

    A. Replace the fuse with the next higher amperage rating.

    B. Check the affected circuit for a short to ground.

    C. Check the affected circuit for an open.

    D. Install a circuit breaker with a smaller amperage rating than the fuse.

TASK A.6

    Answer A is incorrect. Installing a fuse that is above the specified rating could cause major problems on a truck, including a potential electrical fire.

    Answer B is correct. The electrical circuit should be checked for a short to ground when a blown fuse is found.

    Answer C is incorrect. An open circuit will cause the current flow to stop. An open circuit will not cause a fuse to blow.

    Answer D is incorrect. A circuit breaker with a smaller amperage rating is not the correct repair procedure when a blown fuse is found. Sometimes circuit breakers are used during the diagnostic process to prevent the loss of several fuses.

25. Which of the following faults could cause a truck with an automatic transmission to have an inoperative cranking system?

    A. Faulty throttle pedal

    B. Blocked fuel injector

    C. Battery with terminal voltage of 12.6 volts

    D. Open wire at the park/neutral switch

TASK B.11

    Answer A is incorrect. A faulty throttle pedal would not cause an inoperative cranking system. A faulty pedal could cause a low engine power problem.

    Answer B is incorrect. A blocked fuel injector would not cause an inoperative cranking system. A blocked injector would cause a low engine power problem.

    Answer C is incorrect. A battery with 12.6 volts at the terminals is considered to be fully charged and ready for service. This would not cause an inoperative cranking system.

    Answer D is correct. An open park/neutral switch could cause an inoperative cranking system by not providing a path for ground to the starter relay.

TASK A.4

26. Which of the following conditions would be the LEAST LIKELY cause for a blown horn fuse?

  A. Shorted electric horn
  B. Power wire rubbing a metal bracket
  C. Open terminal
  D. Shorted fuse box

Answer A is incorrect. A shorted electric horn would cause increased current flow, which could cause a blown fuse.

Answer B is incorrect. A power wire rubbing a metal bracket could cause increased current flow, which could cause a blown fuse.

Answer C is correct. An open terminal would cause current flow to be interrupted, which would not cause a blown fuse.

Answer D is incorrect. A shorted fuse box could cause increased current flow, which could cause a blown fuse.

TASK C.2

27. A maintenance-free battery is low on electrolyte. Technician A says a defective voltage regulator may cause this problem. Technician B says a loose alternator belt may cause this problem. Who is correct?

  A. A only
  B. B only
  C. Both A and B
  D. Neither A nor B

Answer A is correct. Technician A is correct. A defective voltage regulator can cause overcharging and possible battery boil over.

Answer B is incorrect. A loose alternator belt might cause undercharging, but not overcharging, which is what a low electrolyte level indicates.

Answer C is incorrect. Only Technician A is correct.

Answer D is incorrect. Technician A is correct.

TASK C.3

28. Which of the following conditions could cause a misaligned alternator pulley?

  A. Faulty A/C compressor bearing
  B. Loose alternator mounting bracket
  C. Loose alternator output wire
  D. Faulty water pump bearing

Answer A is incorrect. A faulty A/C compressor bearing could cause noise in the accessory area, but it would not cause a misaligned alternator pulley.

Answer B is correct. A loose alternator mounting bracket could cause the alternator body to move and result in a misaligned pulley.

Answer C is incorrect. A loose alternator output wire could cause the system to undercharge the battery, but it would not cause the pulley to be misaligned.

Answer D is incorrect. A faulty water pump bearing could cause noise in the accessory area, but it would not cause a misaligned alternator pulley.

29. What is the LEAST LIKELY result of a full-fielded alternator?

    A. Low battery voltage due to excessive field current draw
    B. Burned-out light bulbs on the vehicle
    C. Battery boil over
    D. Excessive charging system voltage

TASK C.4

    Answer A is correct. A full-fielded alternator will cause maximum output from the alternator and consequent high battery voltage.

    Answer B is incorrect. A full-fielded alternator can cause excessive output voltage that could lead to burned-out bulbs on the vehicle.

    Answer C is incorrect. Maximum output from an alternator due to full-fielding can cause a battery to boil over due to excessive voltage.

    Answer D is incorrect. A full-fielded alternator will cause excessive battery voltage because the alternator is at maximum output.

30. A heavy-duty truck is being diagnosed for a charging problem. The alternator only charges at 12.2 volts. Technician A says that the voltage drop should be checked on the charging output wire. Technician B says that the voltage drop should be checked on the charging ground circuit. Who is correct?

TASK C.5

    A. A only
    B. B only
    C. Both A and B
    D. Neither A nor B

    Answer A is incorrect. Technician B is also correct.

    Answer B is incorrect. Technician A is also correct.

    Answer C is correct. Both Technicians are correct. Checking the voltage drop in the positive and negative sides of the charging system is always recommended when a charging system is not performing up to specifications.

    Answer D is incorrect. Both Technicians are correct.

31. What is the LEAST LIKELY cause of a discharged battery?

    A. Loose alternator belt
    B. Corroded battery cable connection
    C. Defective starter drive
    D. Parasitic drain

TASK B.3

    Answer A is incorrect. A loose alternator belt could cause a discharged battery by failing to provide the necessary amperage.

    Answer B is incorrect. A corroded battery cable connection could cause a discharged battery by creating an electrical barrier for charging the battery.

    Answer C is correct. A defective starter drive would not likely cause a discharged battery, since it is a mechanical component of the starter assembly.

    Answer D is incorrect. A parasitic drain could cause a discharged battery by pulling amperage out of the battery while the truck is turned off.

TASK C.7

32. Which of the following conditions would most likely cause a burned charging output wire?

    A. Failed alternator rotor assembly
    B. Incorrect alternator pulley
    C. Loose alternator bracket
    D. Voltage regulator stuck at maximum charging mode

    Answer A is incorrect. A failed alternator rotor would cause the alternator to be totally inoperative.

    Answer B is incorrect. An incorrect alternator pulley could cause the charging system to undercharge.

    Answer C is incorrect. A loose alternator bracket could cause belt misalignment.

    Answer D is correct. A voltage regulator that is stuck in the maximum charging mode could cause the charging output wire to get warm and look burned.

TASK D.1

33. The right-side headlight on a truck is very dim, and the left-side headlight is normal. Technician A says that the dimmer switch is likely defective. Technician B says that the left-side headlight could have a bad ground. Who is correct?

    A. A only
    B. B only
    C. Both A and B
    D. Neither A nor B

    Answer A is incorrect. A dimmer switch problem would cause both sides to be affected.

    Answer B is incorrect. A poor ground on the left-side headlight would cause the left side to be affected.

    Answer C is incorrect. Neither Technician is correct.

    Answer D is correct. Neither Technician is correct. A faulty dimmer would affect either the high or low beams on both sides, not just one side. A bad ground on the left side would not likely affect the headlight on the right side.

TASK D.2

34. Technician A says that headlight aim should be checked on a level floor with the vehicle unloaded. Technician B says that some states have very strict laws on adjusting the headlights. Who is correct?

    A. A only
    B. B only
    C. Both A and B
    D. Neither A nor B

    Answer A is incorrect. Technician B is also correct.

    Answer B is incorrect. Technician A is also correct.

    Answer C is correct. Both Technicians are correct. Headlight aim should be checked on a level floor with the vehicle unloaded. Some states have more strict laws than others on headlight aiming procedures.

    Answer D is incorrect. Both Technicians are correct.

35. The dash lights on a medium-duty truck do not work. Technician A says that the fuse to the taillights could be the cause. Technician B says that the rheostat in the headlight switch could be the cause. Who is correct?

TASK D.5

   A. A only

   B. B only

   C. Both A and B

   D. Neither A nor B

Answer A is incorrect. Technician B is also correct.

Answer B is incorrect. Technician A is also correct.

Answer C is correct. Both Technicians are correct. The same fuse that powers the taillights typically provides power for the dash lights. The rheostat in the headlight switch is another component that could possibly cause the dash lights to be inoperative.

Answer D is incorrect. Both Technicians are correct.

36. Technician A says that all International Organization for Standardization (ISO) relays are the same and can be interchanged. Technician B says that terminals 85 and 86 of an ISO relay are connected to the coil inside the relay. Who is correct?

TASK A.8

   A. A only

   B. B only

   C. Both A and B

   D. Neither A nor B

Answer A is incorrect. Only relays with the same part number should be interchanged during the testing sequence of a problem circuit.

Answer B is correct. Only Technician B is correct. Terminals 85 and 86 connect to the coil of an ISO relay. Terminal 30 is typically the supply terminal of the load side of the relay. Terminal 87 is typically the output of the load side of the relay. Terminal 87a is typically the normally closed (NC) contact of the relay.

Answer C is incorrect. Only Technician B is correct.

Answer D is incorrect. Technician B is correct.

37. The right turn signal indicator comes on when the brakes are applied. Technician A says that a bad ground at the right rear stoplight socket could be the cause. Technician B says that a single-filament bulb in a dual-filament socket could be the cause. Who is correct?

TASK D.8

   A. A only

   B. B only

   C. Both A and B

   D. Neither A nor B

Answer A is incorrect. Technician B is also correct.

Answer B is incorrect. Technician A is also correct.

Answer C is correct. Both Technicians are correct. A bad ground in taillight and stoplight circuits can sometimes cause an electrical backfeeding problem that could cause the turn indicator to come on when the brakes are applied. Installing a single-filament bulb in a dual-filament socket could also cause unusual lighting problems.

Answer D is incorrect. Both Technicians are correct.

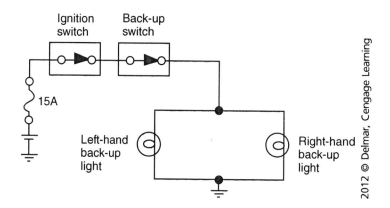

Ignition switch   Back-up switch

15A

Left-hand back-up light   Right-hand back-up light

2012 © Delmar, Cengage Learning

TASK D.10

38. Referring to the figure above, the right-side back-up light circuit is accidentally grounded on the switch side of the bulb in the circuit. Technician A says this condition may blow the back-up light fuse. Technician B says the left-side back-up light may work normally while the right-side back-up light is inoperative. Who is correct?

A. A only

B. B only

C. Both A and B

D. Neither A nor B

Answer A is correct. Only Technician A is correct. A short to ground on the positive side of the bulb will cause increased current flow that will blow the fuse.

Answer B is incorrect. Even if the fuse did not blow, the increased current demand in the circuit on the right side would cause the left-side bulb to be dim. This would happen because the two branches would no longer have equal resistance; rather, the right side would have less resistance and consequently increased current flow.

Answer C is incorrect. Only Technician A is correct.

Answer D is incorrect. Technician A is correct.

TASK E.2

39. Technician A says that a truck with an electronic instrument panel sources its data directly from the sensors on the engine. Technician B states that a data bus-style electronic instrument panel receives information exclusively from the engine control module (ECM). Who is correct?

A. A only

B. B only

C. Both A and B

D. Neither A nor B

Answer A is incorrect. The electronic instrument panel sources information from the data bus via the various chassis electronic control modules (ECMs).

Answer B is incorrect. The electronic instrument panel sources information from other processors as well as the ECM.

Answer C is incorrect. Neither Technician is correct.

Answer D is correct. Neither Technician is correct. A multiplexed instrument cluster is basically a computer that receives inputs from the data bus and converts those inputs to gauge functions. The inputs on the data bus can come from various controllers connected to the bus, such as the ECM, the ABS computer, and the driver information computer.

40. The temperature sending unit needs to be replaced on a heavy-duty truck. Technician A says that this process can be completed without draining the whole cooling system. Technician B says that the engine should be cooled down prior to performing this repair. Who is correct?

TASK E.3

A. A only

B. B only

C. Both A and B

D. Neither A nor B

Answer A is incorrect. Technician B is also correct.

Answer B is incorrect. Technician A is also correct.

Answer C is correct. Both Technicians are correct. The temperature sending unit can be replaced without draining the whole cooling system, but it is advisable to let the engine cool down before beginning the procedure.

Answer D is incorrect. Both Technicians are correct.

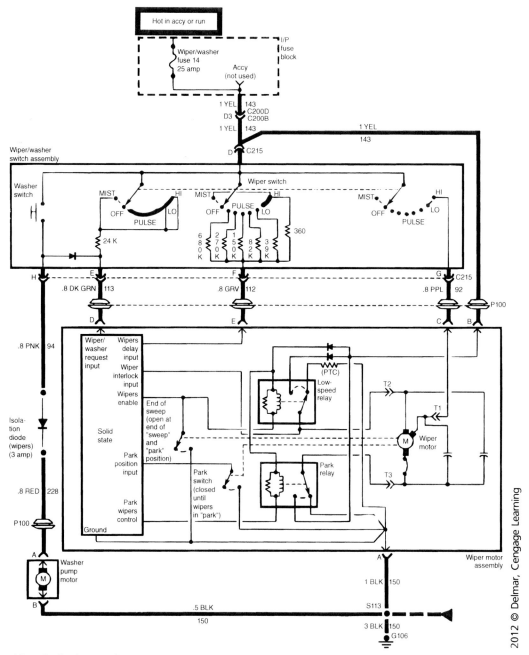

TASK A.3, A.9

41. Referring to the wiring schematic above, Technician A says that the wiper motor receives ground from item P100. Technician B says that Fuse #14 supplies power to the windshield wiper switch. Who is correct?

    A. A only
    B. B only
    C. Both A and B
    D. Neither A nor B

Answer A is incorrect. P100 is a pass-through grommet that allows several wires to be safely routed through the firewall into the cab area.

Answer B is correct. Only Technician B is correct. Fuse #14 is located at the top of the wiring schematic and provides power to the windshield wiper switch.

Answer C is incorrect. Only Technician B is correct.

Answer D is incorrect. Technician B is correct.

42. Which of the following components provides the tachometer signal to the instrument cluster on older medium-duty trucks with diesel engines?

TASK E.5

    A. Primary ignition coil

    B. Secondary ignition coil

    C. Alternator (R terminal)

    D. Vehicle speed sensor

Answer A is incorrect. A diesel engine does not use a primary ignition coil.

Answer B is incorrect. A diesel engine does not use a secondary ignition coil.

Answer C is correct. Some tachometers can derive their signal off the back of the alternator. The tachometer is an electromagnetic gauge, similar to an electronic speedometer, with two coils. It operates from a tachometer signal off the primary ignition coil on gasoline engines. On some diesel engines, the tachometer signal is taken from the alternator R terminal.

Answer D is incorrect. A vehicle speed sensor provides a signal to the speedometer, not the tachometer.

43. A customer says the electric horn on a truck will not turn off. Technician A says the cause could be welded diaphragm contacts inside the horn. Technician B says the relay may be defective. Who is correct?

TASK E.6

    A. A only

    B. B only

    C. Both A and B

    D. Neither A nor B

Answer A is incorrect. Welded diaphragm contacts inside the horn will not allow it to vibrate and sound.

Answer B is correct. Only Technician B is correct. A faulty relay can cause a horn to sound constantly if it is stuck in the energized position.

Answer C is incorrect. Only Technician B is correct.

Answer D is incorrect. Technician B is correct.

44. The windshield washer pump motor runs continuously while the ignition switch is on. Technician A says that the multi-function switch could be shorted. Technician B says that the wiper control module could be defective. Who is correct?

TASK E.10

    A. A only

    B. B only

    C. Both A and B

    D. Neither A nor B

Answer A is incorrect. Technician B is also correct.

Answer B is incorrect. Technician A is also correct.

Answer C is correct. Both Technicians are correct. A shorted multi-function switch could cause a washer motor to run continuously because the wiper switches are incorporated into that switch on several late-model vehicles. A defective wiper control module could cause the washer motor to run continuously on some late-model vehicles.

Answer D is incorrect. Both Technicians are correct.

TASK B.12

45. Technician A says that the starter solenoid pull-in winding can be tested by touching the "start" terminal and the "bat" terminal with the ohmmeter leads. Technician B says that touching the "start" terminal and the solenoid case with the ohmmeter leads will test the starter solenoid hold-in winding. Who is correct?

   A. A only
   B. B only
   C. Both A and B
   D. Neither A nor B

Answer A is incorrect. The pull-in winding can be tested by connecting the ohmmeter leads to the "start" terminal and the "motor" terminal.

Answer B is correct. Only Technician B is correct. The hold-in winding is case grounded to the solenoid case. Connecting the ohmmeter leads to the "start" terminal and the solenoid case is a way of testing the hold-in winding.

Answer C is incorrect. Only Technician B is correct.

Answer D is incorrect. Technician B is correct.

TASK E.12

46. A blower motor will not work on high, but works well on all other speeds. Technician A says that a faulty blower resistor could be the cause. Technician B says that a faulty blower relay could be the cause. Who is correct?

   A. A only
   B. B only
   C. Both A and B
   D. Neither A nor B

Answer A is incorrect. If the blower resistor were bad, then the blower would not work on the lower speeds.

Answer B is correct. Only Technician B is correct. A faulty blower relay would be the likely cause of the blower working on all speeds but high.

Answer C is incorrect. Only Technician B is correct.

Answer D is incorrect. Technician B is correct.

TASK E.13

47. The auxiliary power outlet is inoperative and the fuse is found to be open. What is the most likely cause for this condition?

   A. Loose connection at the power outlet plug
   B. Foreign metal object in the power outlet
   C. Broken wire leading to the power outlet
   D. Open internal connection at the power outlet

Answer A is incorrect. A loose connection would not typically blow the fuse. Such a connection would just cause the components that connect to the outlet to not operate correctly.

Answer B is correct. It is not uncommon for metal objects to fall into the auxiliary power outlets, which can cause a short to ground and overload the fuse.

Answer C is incorrect. A broken wire does not typically cause a fuse to blow.

Answer D is incorrect. An open connection at the power outlet would not cause the fuse to blow.

48. Which of the following conditions would be the LEAST LIKELY cause for all of the power window motors to be inoperative?

TASK E.6

   A. Open internal circuit breaker at the passenger power window motor

   B. Faulty power window master switch

   C. Missing ground at the master switch

   D. Faulty power window circuit breaker

Answer A is correct. An open internal circuit breaker at the passenger power window motor would not cause all of the power window motors to be inoperative. The driver's side power window would still operate.

Answer B is incorrect. A faulty power window master switch could cause all of the power window motors to be inoperative, because the current for all of the windows must flow through the master switch in order to get to ground.

Answer C is incorrect. A missing ground at the master switch would cause all of the power window motors to be inoperative, because this is the only ground supplied to the power window circuit.

Answer D is incorrect. A faulty power window circuit breaker would cause all of the power window motors to be inoperative because they would not receive any voltage.

49. Which of the following is a measurement for electrical pressure?

TASK A.1

   A. Ohm

   B. Amp

   C. Watt

   D. Volt

Answer A is incorrect. The ohm is the unit of electrical resistance.

Answer B is incorrect. The amp is the unit of electrical current flow.

Answer C is incorrect. The watt is the unit of electrical power.

Answer D is correct. The volt is considered the unit of electrical pressure that performs the electrical work in a circuit. The source of voltage on the truck is the battery or the alternator.

50. A truck is being diagnosed for a fuel supply problem. The technician measures the voltage available at the transfer pump terminals and finds 7.5 volts. Technician A says that the problem could be caused by a faulty fuel pump relay. Technician B says that the problem could be caused by a poor ground connection in the supply pump circuit. Who is correct?

TASK E.21

   A. A only

   B. B only

   C. Both A and B

   D. Neither A nor B

Answer A is incorrect. Technician B is also correct.

Answer B is incorrect. Technician A is also correct.

Answer C is correct. Both Technicians are correct. The voltage measurement at the transfer pump terminals reveals a loss of voltage at some point in the fuel system. A faulty fuel pump relay with burned contacts could cause this problem, because some of the voltage would be lost at the bad contacts. A poor ground connection in the supply pump circuit could cause the test results, too. A poor ground will become a point of resistance and will cause a loss of voltage.

Answer D is incorrect. Both Technicians are correct.

PREPARATION EXAM 2—ANSWER KEY

1.	D	18.	B	35.	B
2.	C	19.	B	36.	C
3.	B	20.	C	37.	C
4.	B	21.	B	38.	A
5.	A	22.	D	39.	A
6.	D	23.	D	40.	D
7.	C	24.	B	41.	A
8.	A	25.	A	42.	B
9.	C	26.	B	43.	B
10.	A	27.	C	44.	A
11.	D	28.	A	45.	A
12.	B	29.	C	46.	A
13.	B	30.	D	47.	B
14.	C	31.	C	48.	A
15.	D	32.	B	49.	D
16.	C	33.	C	50.	B
17.	B	34.	D		

PREPARATION EXAM 2—EXPLANATIONS

TASK A.3

1. Technician A says that burned electrical contacts will decrease the electrical resistance in a circuit. Technician B says that an open switch should have continuity. Who is correct?

    A. A only

    B. B only

    C. Both A and B

    D. Neither A nor B

    Answer A is incorrect. Burned electrical contacts will increase electrical resistance in a circuit.

    Answer B is incorrect. An open switch should measure infinite resistance.

    Answer C is incorrect. Neither Technician is correct.

    Answer D is correct. Neither Technician is correct. Burned electrical contacts will increase the resistance in a circuit, which will cause reduced current flow. An open switch will have infinite resistance. A closed switch should have continuity.

2. A truck is in the repair shop with an inoperative power window. During diagnosis, a blown power window fuse is located. Technician A says that a short to ground between the switch and the motor could be the cause. Technician B says that a tight power window motor could be the cause. Who is correct?

TASK A.4

   A. A only
   B. B only
   C. Both A and B
   D. Neither A nor B

Answer A is incorrect. Technician B is also correct.

Answer B is incorrect. Technician A is also correct.

Answer C is correct. Both Technicians are correct. A short to ground on the power feed side of the circuit will very likely blow a fuse due to the current quickly rising because the resistance is very low. A binding or tight power window motor can also cause the current flow to rise quickly and blow the power window fuse. Some manufacturers use circuit breakers in the power window circuits for this reason.

Answer D is incorrect. Both Technicians are correct.

3. A truck is being diagnosed for a problem with the cruise control kicking out when the turn signal is operated. Which of the following would be the most likely cause?

TASK E.19

   A. Faulty throttle pedal
   B. Faulty multi-function switch
   C. Faulty clutch switch
   D. Faulty engine control module (ECM)

Answer A is incorrect. A faulty throttle pedal would cause problems while driving in normal mode.

Answer B is correct. A faulty multi-function switch could cause this problem. The multi-function switch incorporates many switches into one assembly. Some trucks include the cruise control switch on the multi-function switch.

Answer C is incorrect. A faulty clutch switch could cause cruise control problems, but it would not likely cause the cruise to kick out when the turn signal is operated.

Answer D is incorrect. A faulty ECM could cause a number of different problems with a truck. However, the ECM would not likely cause the cruise to kick out when the turn signal is operated.

4. Technician A says that a digital multimeter (DMM) can be used to test current flow directly through the meter in any truck electrical circuit. Technician B states that a current clamp can be used on high-amperage circuits to prevent damage to the meter. Who is correct?

TASK A.2

   A. A only
   B. B only
   C. Both A and B
   D. Neither A nor B

Answer A is incorrect. DMMs typically have a limit on the amount of current that can be measured. This limitation will prevent the use of the internal ammeter on circuits that draw more than approximately 10 amps of current.

Answer B is correct. Only Technician B is correct. Current clamps can be used in conjunction with DMMs to measure higher current levels. These devices connect to the volt/ohm slots of the meter and the clamp is then inserted around the wire that needs to be measured.

Answer C is incorrect. Only Technician B is correct.

Answer D is incorrect. Technician B is correct.

TASK A.2

5. The blower motor in a truck is running very slowly. An ammeter shows a low current flow. Which of the following could cause the described conditions?

    A. High resistance in the circuit
    B. Low resistance in the circuit
    C. Overcharged battery
    D. Shorted blower motor

    Answer A is correct. Increased electrical resistance will cause the blower motor to run slowly while having lower current draw. The current flow is being restricted by the added electrical resistance.

    Answer B is incorrect. Decreased electrical resistance will cause an increase in current flow. If the resistance is greatly reduced, the current can increase to the point of opening the circuit protection device.

    Answer C is incorrect. An overcharged battery would not cause the blower motor to run slower. Higher voltage in the battery would slightly increase the current flow to the blower motor.

    Answer D is incorrect. A shorted blower motor would cause an increase in current flow in the circuit. If the resistance is greatly reduced, the current can increase to the point of opening the circuit protection device.

TASK A.2

6. Technician A says that loose electrical contacts will decrease the electrical resistance in a circuit. Technician B says that an open switch should read very low when using an ohmmeter. Who is correct?

    A. A only
    B. B only
    C. Both A and B
    D. Neither A nor B

    Answer A is incorrect. Loose electrical contacts will increase electrical resistance in a circuit.

    Answer B is incorrect. An open switch should measure out of limits (OL) when using an ohmmeter.

    Answer C is incorrect. Neither Technician is correct.

    Answer D is correct. Neither Technician is correct. Loose electrical contacts will increase the resistance in a circuit, which will cause reduced current flow. An open switch will have infinite resistance. A closed switch should have continuity.

TASK C.1

7. A truck is being diagnosed for a charge indicator light that is not illuminated when the engine is running. The truck has an undercharged battery. Technician A says a blown fuse between the indicator lamp and ignition switch could cause this problem. Technician B says this could be caused by a failed charge light bulb. Who is correct?

    A. A only
    B. B only
    C. Both A and B
    D. Neither A nor B

    Answer A is incorrect. Technician B is also correct.

    Answer B is incorrect. Technician A is also correct.

    Answer C is correct. Both Technicians are correct. A blown fuse will interrupt the current flow to the bulb, causing it to not come on. Also, a failed bulb would not illuminate in the event of a charging system malfunction.

    Answer D is incorrect. Both Technicians are correct.

8. Technician A says that a key-off current draw of two amps could cause the batteries to discharge. Technician B says that two amps is a normal key-off draw for a vehicle with an ECM. Who is correct?

TASK A.5

   A. A only
   B. B only
   C. Both A and B
   D. Neither A nor B

   Answer A is correct. Only Technician A is correct. A two amp key-off draw would cause the batteries to become discharged if present for several hours.

   Answer B is incorrect. Trucks with ECMs will not have large key-off draw. Typically, ECMs only draw a few milliamps when they are powered down.

   Answer C is incorrect. Only Technician A is correct.

   Answer D is incorrect. Technician A is correct.

9. A truck is in the repair shop for a starting problem. There is a clicking sound at the starter when the key is moved to the start position. Technician A says that the starter solenoid could be the cause. Technician B says that worn starter brushes could be the problem. Who is correct?

TASK B.15

   A. A only
   B. B only
   C. Both A and B
   D. Neither A nor B

   Answer A is incorrect. Technician B is also correct.

   Answer B is incorrect. Technician A is also correct.

   Answer C is correct. Both Technicians are correct. A clicking noise while attempting to start an engine could be caused by a faulty starter solenoid or worn starter brushes. A voltage drop test at the solenoid could be performed to further troubleshoot this problem.

   Answer D is incorrect. Both Technicians are correct.

10. Which of the following could result if an air conditioner (A/C) compressor clutch diode fails "open"?

TASK A.7

   A. ECM failure from voltage spikes
   B. Compressor running backward
   C. Clutch coil inoperative due to no current
   D. Clutch coil failure from high current

   Answer A is correct. A failed A/C compressor clutch diode could cause the voltage spike, created when the coil is de-energized, to damage electronic components, such as an ECM.

   Answer B is incorrect. The compressor would not run backward unless the engine was running backward, which is unlikely.

   Answer C is incorrect. The clutch coil would still operate normally if the clutch diode failed. The circuit would not have spike suppression protection if the diode failed.

   Answer D is incorrect. The current in the clutch coil would not be affected if the clutch diode failed "open."

TASK D.7

11. Which of the following problems would most likely cause the stoplights to be inoperative?

  A. Stuck closed lamp back-up switch
  B. Incorrect turn signal flasher
  C. Blown headlight fuse
  D. Blown stoplight fuse

Answer A is incorrect. A stuck closed back-up lamp switch would cause the back-up lights to run continuously any time the key is turned on.

Answer B is incorrect. An incorrect turn signal flasher would cause the turn signals to malfunction. Most turn signal flashers are specific for the number and type of bulbs that are used in the system.

Answer C is incorrect. A blown headlight fuse would cause problems in the headlight operation, but not in the stoplight system.

Answer D is correct. A blown stoplight fuse could cause all of the rear lighting to be inoperative, including the stoplights.

TASK A.8

12. A relay can be tested with a multimeter for all of the following tests EXCEPT:

  A. Resistance of the normally open (NO) contacts
  B. Resistance of the clamping diode
  C. Resistance of the normally closed (NC) contacts
  D. Resistance of the coil

Answer A is incorrect. The NO contacts can be tested on a relay. This reading should be OL on a good relay.

Answer B is correct. It is not possible to test the clamping diode accurately without disassembling the relay.

Answer C is incorrect. The NC contacts can be tested on a relay. This reading should be very low resistance.

Answer D is incorrect. The coil resistance can be tested with an ohmmeter. The reading is typically 60 to 90 ohms on an International Organization for Standardization (ISO) relay.

TASK E.4

13. Which of the following generally activates warning lights and/or warning devices?

  A. Vehicle ignition switch
  B. Closing a switch or sensor
  C. Opening a switch or sensor
  D. Vehicle battery

Answer A is incorrect. The ignition switch only powers up the circuits and is not responsible for switching individual lights on and off.

Answer B is correct. Closing a switch or sensor is usually required to complete a circuit to power up a light or warning device.

Answer C is incorrect. Opening a switch or sensor will interrupt current flow in a circuit, canceling the operation of a light or warning device.

Answer D is incorrect. The vehicle battery only supplies chassis electrical power and cannot switch components on and off.

14. Which of the following tools is the LEAST LIKELY choice for use in diagnosing a problem on the data bus network?

TASK A.10

  A. Oscilloscope
  B. DMM
  C. Continuity tester
  D. Scan tool

Answer A is incorrect. An oscilloscope could be used to view electrical activity on the data bus network.

Answer B is incorrect. A DMM could be used to measure voltage levels on the data bus network.

Answer C is correct. A continuity tester should not be used on data circuits on late-model vehicles.

Answer D is incorrect. A scan tool could be used to retrieve data from the electronic modules that communicate on the data bus network.

15. Technician A says that the battery must be at least 40 percent charged in order for the load test to be valid. Technician B says that the load test must be run for 10 seconds. Who is correct?

TASK B.2

  A. A only
  B. B only
  C. Both A and B
  D. Neither A nor B

Answer A is incorrect. A battery that is 40 percent charged will always fail the load test.

Answer B is incorrect. The battery load test needs to be performed for 15 seconds.

Answer C is incorrect. Neither Technician is correct.

Answer D is correct. Neither Technician is correct. A battery must be at least 75 percent charged in order for a battery load test to be valid. The battery load test must be run for 15 seconds to be valid.

16. Which of the following would be the most likely terminal voltage on a fully charged truck battery?

TASK B.1

  A. 12 volts
  B. 12.2 volts
  C. 12.6 volts
  D. 13.2 volts

Answer A is incorrect. A truck battery with a voltage level of 12 volts is at approximately 25 percent charge.

Answer B is incorrect. A truck battery with a voltage level of 12.2 volts is at approximately 50 percent charge.

Answer C is correct. A fully charged truck battery will have approximately 12.6 volts.

Answer D is incorrect. A battery with 13.2 volts of terminal voltage likely has sulfated cells. A three-minute charge test should be performed to test for this problem.

TASK C.6

17. An alternator is being replaced on a heavy truck. Technician A uses an air tool to install the fastening nut onto the charging output wire. Technician B carefully routes the drive belt around all of the pulleys before releasing the belt tensioner. Who is correct?

    A. A only

    B. B only

    C. Both A and B

    D. Neither A nor B

Answer A is incorrect. An air tool should never be used to install the fastening nut onto the charging output wire. This would likely cause internal damage to the alternator.

Answer B is correct. Only Technician B is correct. The drive belt should be routed around all of the drive pulleys before releasing the tensioner.

Answer C is incorrect. Only Technician B is correct.

Answer D is incorrect. Technician B is correct.

TASK B.6

18. Technician A says that a low battery cannot generate explosive vapors to the extent that a fully charged battery can. Technician B states that it is good practice to wear eye protection when jump-starting a truck. Who is correct?

    A. A only

    B. B only

    C. Both A and B

    D. Neither A nor B

Answer A is incorrect. A low battery is usually caused by excessive cranking. This is when a battery generates most of its explosive vapors.

Answer B is correct. Only Technician B is correct. It is always considered good practice to wear eye protection when working near batteries.

Answer C is incorrect. Only Technician B is correct.

Answer D is incorrect. Technician B is correct.

TASK B.9

19. A truck is being diagnosed for an inoperative starter. A voltage drop test is performed on the solenoid "load side" while the ignition switch is held in the crank mode and 0.1 volts are measured. Technician A says that the solenoid is faulty. Technician B says that this test should only be performed if the battery is at least 75 percent charged. Who is correct?

    A. A only

    B. B only

    C. Both A and B

    D. Neither A nor B

Answer A is incorrect. The voltage drop test revealed only 0.1 volts, which is not excessive.

Answer B is correct. Only Technician B is correct. The truck battery should be at least 75 percent charged when testing the starter circuit.

Answer C is incorrect. Only Technician B is correct.

Answer D is incorrect. Technician B is correct.

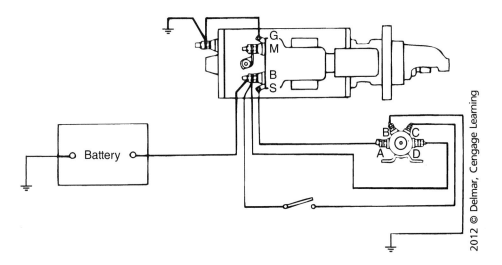

20. Referring to the figure above, which component is being checked if terminals A and D are jumped at the magnetic switch?

TASK B.10

A. Starting switch

B. Battery

C. Magnetic switch

D. Starter

Answer A is incorrect. Terminals C and D would need to be jumped in order to test the starting switch.

Answer B is incorrect. The battery can be tested by monitoring the voltage with a multimeter or by using a battery test tool directly on the battery.

Answer C is correct. The magnetic switch is being by-passed when the jumper is connected from A to D.

Answer D is incorrect. The starter can be tested by monitoring the cranking amperage while the starter is engaged.

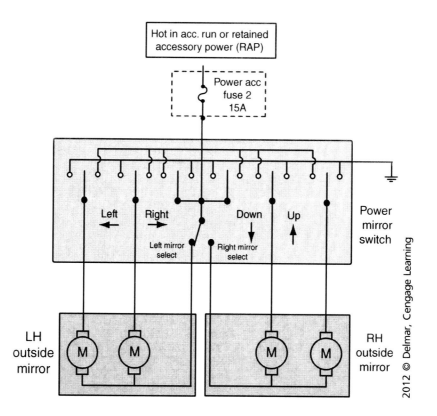

TASK E.11

21. Referring to the figure above, the right-side power mirror functions normally but the left-side power mirror does not function in the up and down directions. Technician A says the mirror select switch could be defective. Technician B says the built-in circuit breaker in the up/down motor could be defective. Who is correct?

A. A only

B. B only

C. Both A and B

D. Neither A nor B

Answer A is incorrect. A defective mirror select switch would prevent both of the mirrors on the left side from working.

Answer B is correct. Only Technician B is correct. A defective circuit breaker in the up/down motor could cause that motor to be inoperative.

Answer C is incorrect. Only Technician B is correct.

Answer D is incorrect. Technician B is correct.

22. Technician A says that a starter drive pinion should not have chamfers on the drive teeth. Technician B states that if the flywheel ring gear is damaged, then the entire flywheel should be replaced. Who is correct?

TASK B.14

  A. A only
  B. B only
  C. Both A and B
  D. Neither A nor B

Answer A is incorrect. A starter drive pinion is machined with a chamfer on the drive teeth to assist in meshing with the flywheel ring gear.

Answer B is incorrect. A flywheel ring gear can be replaced without replacing the entire flywheel.

Answer C is incorrect. Neither Technician is correct.

Answer D is correct. Neither Technician is correct. Starter drive pinions do have chamfered drive teeth to more easily engage the flywheel. Also, the ring gear can be replaced without replacing the whole flywheel.

23. Technician A says that the battery cables only need to be serviced when the starting or charging system is producing problems. Technician B says that battery corrosion only forms on the terminals during cold weather. Who is correct?

TASK B.13

  A. A only
  B. B only
  C. Both A and B
  D. Neither A nor B

Answer A is incorrect. The technician should service battery cables as part of regular maintenance or whenever servicing the battery.

Answer B is incorrect. Battery corrosion can form on terminals in any weather. However, it affects system operation most in cold weather.

Answer C is correct. Neither Technician is correct.

Answer D is correct. Neither Technician is correct. The battery cables should be serviced during maintenance checks or when the battery is being serviced. Battery corrosion can form during all types of weather.

24. Which of the following statements describes a maxi-fuse?

TASK A.6

  A. Higher-quality fuse than standard
  B. Higher-capacity fuse than standard and used in place of fusible links
  C. Another name for a circuit breaker
  D. Found in all truck electrical circuits because of their high current requirements

Answer A is incorrect. Maxi-fuses have higher amperage ratings than standard fuses. These devices are found in power distribution centers on late-model trucks.

Answer B is correct. Maxi-fuses are used on late-model trucks in the place of fusible links. These circuit protection devices are installed in power distribution centers and are easier to test and easier to replace when they open up.

Answer C is incorrect. Maxi-fuses are not considered circuit breakers because they do not reset after they open up due to high current flow.

Answer D is incorrect. Maxi-fuses are typically used in power distribution centers and have higher amp ratings than the smaller fuses.

TASK A.11

25. Technician A says that a scan tool can be used to retrieve trouble codes from an on-board computer of a truck. Technician B says that a multimeter can be used to retrieve trouble codes from a computer of a truck. Who is correct?

    A. A only
    B. B only
    C. Both A and B
    D. Neither A nor B

    Answer A is correct. Only Technician A is correct. A scan tool can be used to retrieve trouble codes from a truck on-board computer. Scan tools can also access live data from the sensors and switches in the system.

    Answer B is incorrect. A multimeter is not typically used to retrieve trouble codes from a truck computer. The scan tool is useful when testing the individual circuits and components after the code has been retrieved.

    Answer C is incorrect. Only Technician A is correct.

    Answer D is incorrect. Technician A is correct.

TASK A.4

26. Technician A says an ammeter should be used to check for a short circuit between circuits. Technician B says to fully charge the battery before checking a circuit for current draw. Who is correct?

    A. A only
    B. B only
    C. Both A and B
    D. Neither A nor B

    Answer A is incorrect. An ohmmeter is the best tool to use to test for shorts between circuits (with the circuit not under power). An ammeter measures current flow only; it cannot pinpoint a cross or short with another circuit.

    Answer B is correct. Only Technician B is correct. Low battery voltage will affect current flow in a circuit, so the battery voltage should always be checked prior to electrical testing.

    Answer C is incorrect. Only Technician B is correct.

    Answer D is incorrect. Technician B is correct.

TASK C.2

27. A truck has a problem of an alternator with zero output. Technician A says the alternator field circuit may have an open circuit. Technician B says the fusible link may be open in the alternator to battery wire. Who is correct?

    A. A only
    B. B only
    C. Both A and B
    D. Neither A nor B

    Answer A is incorrect. Technician B is also correct.

    Answer B is incorrect. Technician A is also correct.

    Answer C is correct. Both Technicians are correct. In an alternator with an open field circuit, there will be zero output due to no magnetic field being present inside the alternator. Also, if the fusible link between the battery and alternator were open, there would be no path for the alternator output to reach the battery.

    Answer D is incorrect. Both Technicians are correct.

28. The accessory drive belt system should be inspected during regular intervals. Technician A says that a serpentine drive belt tensioner should snap back after releasing pressure on it. Technician B says that the drive belt should be replaced at the first sign of cracks on the back side of it. Who is correct?

TASK C.3

  A. A only
  B. B only
  C. Both A and B
  D. Neither A nor B

Answer A is correct. Only Technician A is correct. A drive belt tensioner should have good spring tension when released.

Answer B is incorrect. Small cracks found during a belt inspection should be noted on the repair order. Large cracks and/or missing pieces from the belt are a cause for the belt to be replaced.

Answer C is incorrect. Only Technician A is correct.

Answer D is incorrect. Technician A is correct.

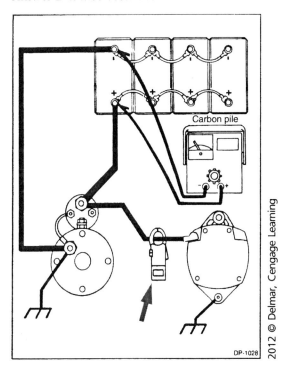

29. Referring to the figure above, what test is being performed with the instrument indicated by the arrow?

TASK C.4

  A. Starter current draw test
  B. Battery load test
  C. Alternator output test
  D. Parasitic battery draw test

Answer A is incorrect. To perform a starter current draw test, the amp clamp would have to be placed around either battery cable.

Answer B is incorrect. Although the battery is being loaded with the carbon pile, the arrow in the picture specifically shows an amp clamp measuring alternator output. The battery is being loaded to force maximum output from the alternator.

Answer C is correct. The arrow shows an amp clamp being used to measure the current output through the alternator output wire.

Answer D is incorrect. Parasitic battery drain tests are not done using an amp clamp. A much more sensitive current measuring device, such as a DMM, is needed for this test.

TASK C.5

30. What is the most likely test that could prove the presence of a poor connection at the charging insulated circuit?

    A. Voltage drop test of the charging ground circuit
    B. Voltage drop test of the negative battery cable
    C. Voltage drop test of the positive battery cable
    D. Voltage drop test of the charging output wire

    Answer A is incorrect. The charging ground is not considered the insulated side of the charging circuit.

    Answer B is incorrect. Testing the negative battery cable is most effective when done while cranking the engine.

    Answer C is incorrect. Testing the positive battery cable is most effective when done while cranking the engine.

    Answer D is correct. Performing a voltage drop test in the charging output wire will test the quality of the wire and associated connections. This test should be done while the engine is running with some electrical loads turned on.

TASK B.4

31. Technician A says that battery acid spills should be cleaned up when found in order to prevent major corrosion in the battery box. Technician B says that water and baking soda should be used to neutralize battery acid if spilled. Who is correct?

    A. A only
    B. B only
    C. Both A and B
    D. Neither A nor B

    Answer A is incorrect. Technician B is also correct.

    Answer B is incorrect. Technician A is also correct.

    Answer C is correct. Both Technicians are correct. Battery acid should always be cleaned up quickly to prevent major corrosion in the battery box, as well as other parts of the truck. Baking soda and water is a good combination to use to neutralize spilled battery acid.

    Answer D is incorrect. Both Technicians are correct.

TASK C.7

32. All of the following wire repair methods could be used on the charging circuit EXCEPT:

    A. Water-resistant harness replacement
    B. Crimp-style butt connectors
    C. Crimp-and-seal connectors
    D. Solder and heat shrink

    Answer A is incorrect. Using a replacement harness that is water resistant is an acceptable repair.

    Answer B is correct. Crimp-style butt connectors are not a recommended repair practice due to the lack of water resistance. This method will not seal out water from the charging circuit, which would cause a future corrosion problem.

    Answer C is incorrect. Using crimp-and-seal connectors is an acceptable repair on the charging circuit.

    Answer D is incorrect. Using solder and heat shrink is an acceptable repair practice on the charging system circuit.

33. All of the following switches could be incorporated into the multi-function switch EXCEPT:

TASK D.3,
D.7, D.9

A. Dimmer switch

B. Turn signal switch

C. Stoplight switch

D. Headlight switch

Answer A is incorrect. Many multi-function switches incorporate a dimmer switch.

Answer B is incorrect. Many multi-function switches incorporate a turn signal switch.

Answer C is correct. The stoplight switch is never incorporated into the multi-function switch.

Answer D is incorrect. Many multi-function switches incorporate a headlight switch.

34. Technician A says that a corroded parking lamp socket could cause a brighter than normal lamp assembly. Technician B says that a single-filament bulb can be used in place of a dual-filament bulb. Who is correct?

TASK D.4

A. A only

B. B only

C. Both A and B

D. Neither A nor B

Answer A is incorrect. A corroded parking lamp socket would add electrical resistance, which would reduce the brightness of the bulb.

Answer B is incorrect. Putting a single-filament bulb in a dual-filament socket will cause lighting problems.

Answer C is incorrect. Neither Technician is correct.

Answer D is correct Neither Technician is correct. Corrosion would cause extra electrical resistance, which would cause the lamps to be less bright than normal. If a socket is made to fit a dual-filament bulb, then a dual-filament bulb must be used as a replacement. If a single-filament bulb is installed, then the circuit will likely not work correctly.

35. Which of the following components could be used to progressively dim dash lights?

TASK D.5

A. Voltage limiter

B. Rheostat

C. Transistor

D. Diode

Answer A is incorrect. A voltage limiter is used to regulate power to instrument gauges at a constant voltage level. There is no way to adjust the voltage output with one of these.

Answer B is correct. A rheostat is a variable resistor. By turning the rheostat, the resistance is either increased or decreased, thereby changing the voltage and current flow to the bulb and altering its brightness.

Answer C is incorrect. A transistor is a three-terminal, electronic switching device.

Answer D is incorrect. A diode is simply a one-way electrical check valve that allows current flow in one direction but not the other. It will not affect the resistance of the circuit in the direction of current flow.

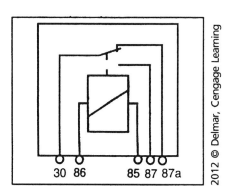

TASK A.9

36. Referring to the figure above, each pin in a 5-pin mini-relay is identified with a number. What is pin #30 for?

    A. Control power in

    B. Control ground

    C. High-amperage power in

    D. High-amperage power out, normally closed

    Answer A is incorrect. Control power in is pin #86.

    Answer B is incorrect. Control ground is pin #85.

    Answer C is correct. Pin #30 is typically used for high-amperage power in. Pin #30 is on the load side of the relay and controls the high current flow. On a bench test, pin #30 should have continuity with pin #87a and should be OL with pin #87.

    Answer D is incorrect. Normally closed power out is pin #87a.

TASK D.10

37. The back-up alarm is inoperative, but the back-up lamps work as designed. Technician A says that an open alarm relay could be the cause. Technician B says that a blown inline fuse to the alarm could be the cause. Who is correct?

    A. A only

    B. B only

    C. Both A and B

    D. Neither A nor B

    Answer A is incorrect. Technician B is also correct.

    Answer B is incorrect. Technician A is also correct.

    Answer C is correct. Both Technicians are correct. A bad relay or a blown fuse could cause the back-up alarm to be inoperative. The back-up lights work, so the back-up switch is operating as designed.

    Answer D is incorrect. Both Technicians are correct.

38. The taillights on the tractor burn normally but the taillights on the trailer are dim. Technician A says that the light cord could have too much resistance. Technician B says that the tractor taillight bulbs could be open. Who is correct?

TASK D.11

A. A only

B. B only

C. Both A and B

D. Neither A nor B

Answer A is correct. Only Technician A is correct. A light cord that has resistance could cause the trailer taillights to be dim because voltage will be dropped in the cord. The loss of voltage will cause the trailer lights to be dim.

Answer B is incorrect. Open taillight bulbs would cause the tractor lights to be inoperative.

Answer C is incorrect. Only Technician A is correct.

Answer D is incorrect. Technician A is correct.

39. The oil pressure gauge intermittently moves to the area above the high setting. Technician A says that the oil pressure should be checked with a manual gauge to verify oil pressure. Technician B says that an open wire in the oil gauge could be the cause of this problem. Who is correct?

TASK E.1

A. A only

B. B only

C. Both A and B

D. Neither A nor B

Answer A is correct. Only Technician A is correct. It would be wise to verify the oil pressure with a manual gauge whenever diagnosing an oil pressure gauge concern.

Answer B is incorrect. An open wire would not cause an intermittent problem.

Answer C is incorrect. Only Technician A is correct.

Answer D is incorrect. Technician A is correct.

40. Which of the following devices is LEAST LIKELY to be used as an input to an electronic gauge assembly?

TASK E.2

A. Thermistor

B. Piezo resistor

C. Body control module (BCM)

D. Light-emitting diode (LED)

Answer A is incorrect. A thermistor is a typical input to a temperature gauge used in an electronic gauge assembly.

Answer B is incorrect. A piezo resistor is a typical input to an oil pressure gauge used in an electronic gauge assembly.

Answer C is incorrect. The BCM is sometimes used as an input to an electronic gauge assembly.

Answer D is correct. An LED would be considered an output device because it is a type of lamp.

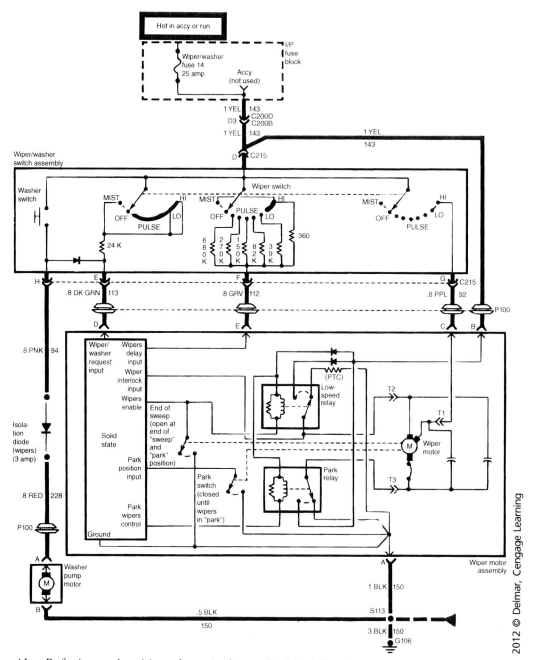

TASK A.9

41. Referring to the wiring schematic above, all of the following statements are correct EXCEPT:

  A. The wiper switch is an input to the BCM.

  B. The windshield wiper circuit is grounded at G106.

  C. P100 is a pass-through grommet for the windshield wiper circuit.

  D. The wiper motor has two speeds.

  Answer A is correct. The windshield wiper switch does not connect to a BCM. The switch directly connects to the wiper motor assembly.

  Answer B is incorrect. G106 is located at the bottom of the wiring schematic. It provides the ground for the windshield wiper circuit.

  Answer C is incorrect. P100 is a pass-through grommet for the wires of the windshield wiper circuit.

  Answer D is incorrect. The wiring schematic does show that the windshield wiper motor has two speeds. The schematic also shows that the windshield wiper system has a wiper delay function.

42. The speedometer in an electronically managed truck is not accurate. Which of the following is the LEAST LIKELY problem?

TASK E.5

    A. The rear axle ratio has not been correctly programmed to the engine computer.

    B. The transmission speed sensor has not been calibrated.

    C. The tire rolling radius has not been correctly programmed into the engine computer.

    D. The engine computer was not reprogrammed when new rear tires were installed.

Answer A is incorrect. If the rear axle ratio has not been properly programmed into the engine computer, the speedometer readings will be inaccurate.

Answer B is correct. The transmission speed sensor does not need to be calibrated; it signals the engine computer output shaft speed.

Answer C is incorrect. If the tire rolling radius has not been properly programmed to the engine computer, incorrect speedometer readings will result.

Answer D is incorrect. Some engine management systems monitor tire wear. If new tires are installed and the engine computer is not reprogrammed, false speedometer readings will result.

43. A truck with dual electric horns is being diagnosed. Technician A says that the horns will likely be wired in series with each other. Technician B says that these systems typically use a high-note horn and a low-note horn. Who is correct?

TASK E.7

    A. A only

    B. B only

    C. C. Both A and B

    D. Neither A nor B

Answer A is incorrect. Dual electric horns are usually wired in parallel with each other in order for each one to receive full battery voltage when operated.

Answer B is correct. Only Technician B is correct. Dual electric horns typically include a high-note horn and a low-note horn to create a louder sound.

Answer C is incorrect. Only Technician B is correct.

Answer D is incorrect. Technician B is correct.

44. Technician A says a binding mechanical wiper linkage can result in no wiper operation. Technician B says a control circuit shorted to ground can cause constant wiper operation. Who is correct?

TASK E.9

    A. A only

    B. B only

    C. Both A and B

    D. Neither A nor B

Answer A is correct. Only Technician A is correct. Binding wiper linkage can result in no wiper operation.

Answer B is incorrect. A shorted control circuit should blow the fuse, not cause it to operate constantly.

Answer C is incorrect. Only Technician A is correct.

Answer D is incorrect. Technician A is correct.

45. Which of the following components is found in a truck starting circuit?

    A. Solenoid
    B. Ballast resistor
    C. Voltage regulator
    D. Engine control module (ECM)

Answer A is correct. A solenoid is typically used on top of a starter to engage the pinion to the flywheel and make the high-current connection between the battery and starter motor.

Answer B is incorrect. A ballast resistor is typically used only in some spark ignition systems.

Answer C is incorrect. A voltage regulator is part of the charging system.

Answer D is incorrect. The ECM is normally not involved in starting an engine. The exception to this statement would be trucks equipped with auto-start systems.

46. Technician A states the auxiliary power outlet is powered at all times. Technician B states that an auxiliary power outlet can be used to jump start another truck. Who is correct?

    A. A only
    B. B only
    C. Both A and B
    D. Neither A nor B

Answer A is correct. Only Technician A is correct. Most truck manufacturers wire the auxiliary power outlet to be hot at all times.

Answer B is incorrect. The auxiliary power outlet should not be used to jump start another truck. This outlet is available for connecting low power accessory devices to the truck's 12 volt system.

Answer C is incorrect. Only Technician A is correct.

Answer D is incorrect. Technician A is correct.

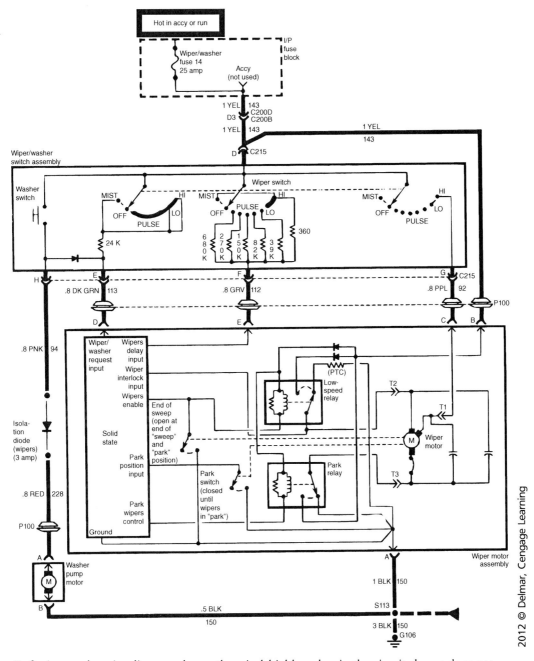

TASK A.9,
E.10

47.   Referring to the wire diagram above, the windshield washer in the circuit shown does not
      operate. The wiper motor operates normally. Technician A says the wiper/washer fuse may
      be open. Technician B says the isolation diode may have an open circuit. Who is correct?

      A.   A only
      B.   B only
      C.   Both A and B
      D.   Neither A nor B

      Answer A is incorrect. A blown wiper/washer fuse would cause all wiper operations to be
      inoperative.

      Answer B is correct. Only Technician B is correct. An open isolation diode would cause the
      washer pump to be inoperative, but the other wiper functions would be normal.

      Answer C is incorrect. Only Technician B is correct.

      Answer D is incorrect. Technician B is correct.

TASK E.18

48. Which of the following steps would most likely be performed when replacing the passenger-side power door lock switch?

    A. Disconnect the clip from the lock switch.
    B. Remove the key lock tumbler.
    C. Remove the BCM.
    D. Remove the door panel.

    Answer A is correct. The clip would need to be disconnected from the passenger lock switch in order to remove it from the truck.

    Answer B is incorrect. The key lock tumbler would not have to be removed to replace the passenger-side power door lock switch.

    Answer C is incorrect. The BCM would not have to be removed to replace the passenger-side power door lock switch.

    Answer D is incorrect. The door panel would not have to be removed to replace the passenger-side power door lock switch.

TASK A.2

49. When using a voltmeter to perform a voltage drop test in a circuit, the leads should be connected in what way?

    A. To the battery terminals
    B. From the positive battery terminal to ground
    C. In series with the circuit being tested
    D. In parallel with the circuit being tested

    Answer A is incorrect. The leads of the voltmeter would be connected to the battery terminals while testing for battery "open circuit" voltage or while testing the charging voltage level. This would not be done when testing voltage drops in other circuits.

    Answer B is incorrect. The leads of the voltmeter would be connected to the positive battery terminal and ground when checking battery voltage.

    Answer C is incorrect. A voltmeter is not usually connected in series with the circuit being tested. Ammeters are connected in series with electrical circuits in order to measure current flow.

    Answer D is correct. The voltmeter leads are connected in parallel with the circuit or component being tested. This test can be performed on electrical components while the circuit is energized.

TASK A.12

50. Which of the following engine parameters would LEAST LIKELY be present in the data list while using a scan tool?

    A. Throttle position
    B. Speed sensor resistance
    C. Turbo boost pressure
    D. Engine coolant temperature

    Answer A is incorrect. Throttle position data is usually available in the data list while using a scan tool on an engine computer.

    Answer B is correct. The speed sensor resistance is not typically available in the data list of an engine computer. This data would have to be manually measured with a digital ohmmeter by the technician.

    Answer C is incorrect. Turbo boost data is usually available in the data list while using a scan tool on an engine computer.

    Answer D is incorrect. Engine coolant temperature data is usually available in the data list while using a scan tool on an engine computer.

PREPARATION EXAM 3—ANSWER KEY

1.	A	18.	C	35.	D
2.	A	19.	B	36.	D
3.	D	20.	B	37.	C
4.	C	21.	A	38.	B
5.	B	22.	C	39.	C
6.	B	23.	B	40.	D
7.	B	24.	C	41.	B
8.	B	25.	C	42.	C
9.	A	26.	D	43.	C
10.	A	27.	B	44.	C
11.	D	28.	A	45.	C
12.	D	29.	A	46.	C
13.	A	30.	B	47.	C
14.	C	31.	C	48.	B
15.	A	32.	B	49.	B
16.	D	33.	D	50.	D
17.	A	34.	D		

PREPARATION EXAM 3—EXPLANATIONS

1. Technician A says that an electrical switch that has continuity will allow current to flow when the switch is closed. Technician B says that a piece of wire that has high resistance will have increased current flow. Who is correct?

TASK A.3

   A. A only
   B. B only
   C. Both A and B
   D. Neither A nor B

   Answer A is correct. Only Technician A is correct. A closed switch should allow current to flow.

   Answer B is incorrect. A wire with added resistance will have reduced current flow.

   Answer C is incorrect. Only Technician A is correct.

   Answer D is incorrect. Technician A is correct.

TASK A.4

2. Technician A says that a voltmeter could be used to find an open circuit by grounding the black lead and using the red lead to test for voltage available throughout the circuit. Technician B says that an open circuit would create a very high current flow and likely would blow a fuse. Who is correct?

    A.   A only
    B.   B only
    C.   Both A and B
    D.   Neither A nor B

    Answer A is correct. Only Technician A is correct. A voltmeter used this way would be a very useful tool to use for finding an open circuit. Care must be taken to not damage the circuit in any way while testing for voltage. Never damage the insulation or connectors when testing electrical circuits. Small T-pins can be used to back-probe connectors to allow the voltage tests to be done.

    Answer B is incorrect. Open circuits do not allow any amperage to flow, so the circuit protection devices would not be affected.

    Answer C is incorrect. Only Technician A is correct.

    Answer D is incorrect. Technician A is correct.

TASK E.17

3. Which of the following conditions could cause the power locks to operate intermittently?

    A.   Blown power lock fuse
    B.   Open power lock circuit breaker
    C.   Broken wire near the power lock actuator
    D.   Chaffed wire near the power lock switch

    Answer A is incorrect. A blown power lock fuse would cause the power locks to be totally inoperative.

    Answer B is incorrect. An open power lock circuit breaker would cause the power locks to be totally inoperative.

    Answer C is incorrect. A broken wire near the power lock actuator would cause the power locks to be totally inoperative.

    Answer D is correct. A chaffed wire near the power door lock switch could cause intermittent operation of the power lock system.

TASK A.3

4. Technician A says that an open switch should have infinite resistance. Technician B says that a closed switch should have continuity. Who is correct?

    A.   A only
    B.   B only
    C.   Both A and B
    D.   Neither A nor B

    Answer A is incorrect. Technician B is also correct.

    Answer B is incorrect. Technician A is also correct.

    Answer C is correct. Both Technicians are correct. An open switch would not provide any path for current to flow. An ohmmeter would measure infinite resistance when used to test an open switch. A closed switch does allow a path for current to flow. An ohmmeter would measure nearly zero ohms of resistance when used to test a closed switch. A continuity tester would show continuity when used to test a closed switch.

    Answer D is incorrect. Both Technicians are correct.

5. A truck/trailer combination comes in with a dim left-rear taillight on the trailer. Technician A says that this may be due to a corroded pin on the left side of the trailer connector. Technician B states that the problem may be caused by a poor ground connection at the left-rear taillight. Who is correct?

TASK A.4

A. A only

B. B only

C. Both A and B

D. Neither A nor B

Answer A is incorrect. The taillight circuit uses only one pin in the trailer connector. Therefore, a faulty pin would have to affect the taillights on both sides equally.

Answer B is correct. Only Technician B is correct. A poor ground connection at the left-rear trailer taillight may cause excessive resistance and low current flow, therefore causing a dim light.

Answer C is incorrect. Only Technician B is correct.

Answer D is incorrect. Technician B is correct.

6. Which tool would be LEAST LIKELY used when checking for key-off battery drain problems?

TASK A.5

A. Low amp probe

B. Ohmmeter

C. Ammeter

D. Amp clamp

Answer A is incorrect. A low amp probe is an amp clamp that will measure low levels of current flow. This tool is an excellent choice for measuring key-off draw.

Answer B is correct. An ohmmeter would not be the correct tool to use when measuring key-off draw on the battery.

Answer C is incorrect. An ammeter is commonly used to measure key-off draw on batteries. This tool must be connected in series with the battery cable in order to work.

Answer D is incorrect. An amp clamp is sometimes accurate enough to measure key-off draw on a battery. This tool is connected to the circuit by clamping the wide jaws around the wire through which the electricity that needs to be measured is flowing.

7. Which of the following is part of a truck charging system?

TASK C.2

A. Voltage solenoid

B. Voltage regulator

C. Voltage transducer

D. Magnetic switch

Answer A is incorrect. There is no voltage solenoid in a truck charging system.

Answer B is correct. A voltage regulator is an integral part of the charging system, whether it is external or internally mounted in the alternator.

Answer C is incorrect. There is no voltage transducer in a truck charging system.

Answer D is incorrect. A magnetic switch is part of the cranking circuit, not the charging circuit.

TASK A.7

8. Which of the following devices would most likely be used as a spike suppression device for an electromagnetic coil?

    A. Relay

    B. Diode

    C. Transistor

    D. Thermistor

Answer A is incorrect. A relay is an electrical device that is used to supply current to electrical loads. Relays allow a large current circuit to be controlled by a small current circuit.

Answer B is correct. A diode is often used as a spike suppression device for electromagnetic coils. Resistors are sometimes used in place of diodes.

Answer C is incorrect. A transistor is a solid-state electronic component that is used as a switch or as a current multiplier in control modules.

Answer D is incorrect. A thermistor is an electronic device that changes resistance when the temperature changes. These devices are used as temperature sensors for electronic circuits.

TASK B.15

9. A truck is being diagnosed for excessive starter noise while the engine is cranking. Technician A says that the starter may have incorrect clearance at the drive gear to ring gear. Technician B says that the starter mounting bolts may be overtightened. Who is correct?

    A. A only

    B. B only

    C. Both A and B

    D. Neither A nor B

Answer A is correct. Only Technician A is correct. Incorrect clearance between the starter drive and the ring gear could cause unusual noise to occur while cranking the engine.

Answer B is incorrect. Starter mounting bolts that are too tight would not likely cause excessive starter noise while cranking the engine.

Answer C is incorrect. Only Technician A is correct.

Answer D is incorrect. Technician A is correct.

TASK A.8

10. What is the purpose of the clamping diode that connects in parallel with the coil of the relay?

    A. Controls voltage spikes as the relay is de-energized

    B. Controls voltage spikes as the relay is energized

    C. Assists in creating the magnetic field when the relay is energized

    D. Assists in creating the magnetic field when the relay is de-energized

Answer A is correct. A clamping diode is used to suppress the voltage spike that is created as the relay is de-energized.

Answer B is incorrect. A voltage spike is not created when the relay is energized.

Answer C is incorrect. The clamping diode does not have electrical flow when the relay is energized. It only allows electrical flow when the voltage spike is produced when the relay is de-energized.

Answer D is incorrect. The clamping diode does not have electrical flow when the relay is energized. It only allows electrical flow when the voltage spike is produced when the relay is de-energized.

11. Technician A says that all stoplight switches are air activated. Technician B states that stoplight switches route current directly to the stoplights. Who is correct?

    A. A only

    B. B only

    C. Both A and B

    D. Neither A nor B

TASK D.7

Answer A is incorrect. A stoplight switch is usually located on the brake pedal on medium-duty trucks with hydraulic brakes.

Answer B is incorrect. The current from a stoplight switch is most often routed through the turn signal switch to allow for proper operation of the brake lights and turn signals simultaneously.

Answer C is incorrect. Neither Technician is correct.

Answer D is correct. Neither Technician is correct. Trucks with hydraulic brakes usually have a brake switch that is activated by moving the brake pedal. Also, it is common to route the current from the brake switch through the turn signal switch in order to work in conjunction with the turn signals on a two-bulb system.

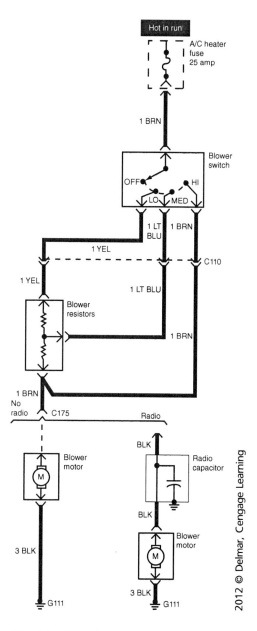

TASK A.9

12. All of the following statements about the wiring schematic above are correct EXCEPT:

  A. The blower motor has three speeds.
  B. The blower switch by-passes the blower resistors when "HI" speed is selected.
  C. The A/C heater fuse provides power for the blower motor circuit.
  D. C175 provides ground for the blower motor.

Answer A is incorrect. The schematic shows that the blower motor has low, medium, and high speeds.

Answer B is incorrect. The blower motor receives full voltage when HI is selected on the blower switch.

Answer C is incorrect. The source for voltage in the blower circuit is the A/C heater fuse, located at the top of the schematic.

Answer D is correct. G111 provides the ground for the blower motor. G111 is located at the bottom of the schematic.

13. The ground circuit for the fuel sender has failed. Technician A says that only the fuel gauge will be affected. Technician B states that all of the gauges will be affected because the gauges share a common ground. Who is correct?

TASK E.3

   A. A only
   B. B only
   C. Both A and B
   D. Neither A nor B

   Answer A is correct. Only Technician A is correct. The ground for the fuel-level gauge is independent of other grounds in the instrumentation circuit. Therefore, only the fuel gauge will be affected.

   Answer B is incorrect. All the gauges shown should be independently grounded.

   Answer C is incorrect. Only Technician A is correct.

   Answer D is incorrect. Technician A is correct.

14. Technician A says that the J1939 data bus uses two 120 ohm terminating resistors. Technician B says that the J1939 data bus uses a two-wire design. Who is correct?

TASK A.10

   A. A only
   B. B only
   C. Both A and B
   D. Neither A nor B

   Answer A is incorrect. Technician B is also correct.

   Answer B is incorrect. Technician A is also correct.

   Answer C is correct. Both Technicians are correct. The J1939 data bus uses two 120 ohm terminating resistors that are connected in parallel with the two data wires of the system. The resistance of the network should be 60 ohms when tested with an ohmmeter at the data connector. This test should be done with the batteries disconnected. The J1939 data bus consists of two data wires. Some networks have a shield wire that reduces electrical interference, but this wire does not transmit any data.

   Answer D is incorrect. Both Technicians are correct.

15. All of the following are acceptable battery and cable maintenance procedures EXCEPT:

   A. Remove the negative battery cable last and reinstall first to avoid sparks.
   B. Clean corrosion and moisture accumulation on the battery top with a water and baking soda solution.

TASK B.3

   C. Only replace battery cable ends with proper solder or crimp-on terminals and heat-shrink tubing.
   D. Coat battery terminal ends with a protective grease to retard corrosion.

   Answer A is correct. The negative cable should be removed first and connected last to reduce the likelihood of creating an electrical spark if the wrench touches a metallic surface while touching the battery terminals.

   Answer B is incorrect. Water and baking soda can be used to safely clean the battery case. This substance helps neutralize the acidic chemicals on and around the battery.

   Answer C is incorrect. Permanent battery terminal repairs should be made with crimp-on terminals. Heat-shrink tubing is also recommended as a way to reduce moisture intrusion.

   Answer D is incorrect. The battery terminal ends should be coated with protective grease to reduce corrosion. This grease should be applied after the connections are secure.

TASK B.5

16. A technician is attempting to charge a battery and yet, according to the ammeter on the charge, it will not accept a charge. What is the LEAST LIKELY source of the problem?

   A. The battery is already fully charged.

   B. The battery is highly sulfated.

   C. Poor contact exists between the charging clamp and the battery post.

   D. Excessive moisture accumulation has caused surface discharge.

   Answer A is incorrect. A fully charged battery will typically resist being further charged.

   Answer B is incorrect. A highly sulfated battery will sometimes resist a charge from a battery charger.

   Answer C is incorrect. A poor connection at the battery charging clamp could cause the battery to not accept a charge.

   Answer D is correct. Moisture on the battery would not typically cause a battery to not accept a charge.

TASK C.6

17. Technician A says that some alternators are mounted using a pad-type bracket. Technician B says that some alternators are mounted using rivets. Who is correct?

   A. A only

   B. B only

   C. Both A and B

   D. Neither A nor B

   Answer A is correct. Only Technician A is correct. Many alternators are mounted using a pad-type bracket.

   Answer B is incorrect. Alternators are typically mounted with threaded fasteners that should be tightened to the correct torque during installation.

   Answer C is incorrect. Only Technician A is correct.

   Answer D is incorrect. Technician A is correct.

TASK B.7

18. The low voltage disconnect (LVD) system opens (turns off power) when the battery voltage drops to what level?

   A. 8.4 volts

   B. 12.6 volts

   C. 10.4 volts

   D. 9.6 volts

   Answer A is incorrect. A battery voltage level of 8.4 would be too low to allow the engine to start.

   Answer B is incorrect. A battery voltage level of 12.6 is the fully charged voltage on truck batteries.

   Answer C is correct. Most LVD systems will open when the battery voltage drops to 10.4 volts. This is the factory preset level and can be adjusted to suit individual needs.

   Answer D is incorrect. A battery voltage level of 9.6 would be too low to allow the engine to start. The specification for performing a battery load test is 9.6 volts.

19. Which of the following problems would be the LEAST LIKELY cause for an excessive voltage drop in the battery's positive cable?

    A. Corrosion in the battery cable
    B. Overcharged batteries
    C. Undersized positive battery cable
    D. Frayed wires in the battery cable

TASK B.9

Answer A is incorrect. Corrosion in the battery cable could contribute to excessive voltage drop in the battery cable. Corrosion increases the resistance in the battery cable.

Answer B is correct. Overcharged batteries would not cause excessive voltage drop in the battery cable. Technicians should closely monitor all batteries while charging them to prevent an overcharged condition.

Answer C is incorrect. A battery cable that is too small will cause excessive voltage drop in the battery cable. Care should be taken to always use the correct size replacement wires when repairing trucks.

Answer D is incorrect. Frayed wires in the battery cable can cause excessive voltage drop. This condition reduces the cable's ability to carry the large amount of amperage needed during engine cranking.

20. All of the following components are parts of the starter control circuit EXCEPT:

    A. Park/neutral switch
    B. Positive battery cable
    C. Starter relay
    D. Ignition switch

TASK B.10

Answer A is incorrect. The park/neutral switch is in the starter control circuit to prevent the starter from operating when the vehicle is in gear.

Answer B is correct. The positive battery cable is in the starter's high current/load circuit.

Answer C is incorrect. The starter relay is in the starter control circuit to limit the amount of current flowing through the ignition switch.

Answer D is incorrect. The ignition switch is in the starter control circuit to allow the driver to initiate the start sequence.

21. The wipers on a truck equipped with electric windshield wipers will not park. Technician A says the activation arm for the park switch is broken or out of adjustment. Technician B says a defective wiper switch will cause this condition. Who is correct?

    A. A only
    B. B only
    C. Both A and B
    D. Neither A nor B

TASK E.9

Answer A is correct. Only Technician A is correct. The wipers will not park correctly if the activation arm for the park switch is broken or out of adjustment.

Answer B is incorrect. A faulty wiper switch only controls the on-off functions of the motor, not the park position.

Answer C is incorrect. Only Technician A is correct.

Answer D is incorrect. Technician A is correct.

TASK B.14

22. Technician A says that the replacement starter assembly should be inspected carefully prior to its installation onto the engine. Technician B says that the replacement starter should be bench tested prior to installing it onto the engine. Who is correct?

    A. A only
    B. B only
    C. Both A and B
    D. Neither A nor B

    Answer A is incorrect. Technician B is also correct.

    Answer B is incorrect. Technician A is also correct

    Answer C is correct. Both Technicians are correct. All replacement starters should be carefully inspected, as well as bench tested, prior to installation onto the engine.

    Answer D is incorrect. Both Technicians are correct.

TASK B.13

23. Technician A says that all replacement batteries should be fast charged for 15 minutes prior to their installation into the truck. Technician B says that the battery cable terminals should be cleaned and protected when installing a replacement battery. Who is correct?

    A. A only
    B. B only
    C. Both A and B
    D. Neither A nor B

    Answer A is incorrect. A replacement battery should not have to be charged prior to installing it into the truck. However, it is a good practice to check the terminal voltage to assure that the battery is ready for installation.

    Answer B is correct. Only Technician B is correct. It is a good practice to clean and protect the battery cable terminals when replacing the battery.

    Answer C is incorrect. Only Technician B is correct.

    Answer D is incorrect. Technician B is correct.

TASK A.8

24. Which of the following would most likely be the resistance test result of a good International Organization for Standardization (ISO) relay?

    A. 1 ohm when connected to terminals 30 and 85
    B. Out of limits (OL) when connected to terminals 30 and 87a
    C. 80 ohms when connected to terminals 85 and 86
    D. 1 ohm when connected to terminals 30 and 87

    Answer A is incorrect. The ohmmeter reading for terminals 30 and 85 should be out of limits (OL). Terminal 30 is on the "load" side of the relay, and terminal 85 is on the "coil" side of the relay.

    Answer B is incorrect. The ohmmeter reading for terminals 30 and 87a should be near zero because they are the normally closed (NC) contacts of the relay.

    Answer C is correct. Terminals 85 and 86 connect to the coil of this type of relay and the measurement should be approximately 60 to 90 ohms.

    Answer D is incorrect. The ohmmeter reading for terminals 30 and 87 should be out of limits (OL). These two terminals are the normally open (NO) contacts of the relay.

25. Technician A says that the main data link connector (DLC) is a round 9-pin connector. Technician B says that flash code diagnostics can be used to retrieve DTCs from the truck. Who is correct?

TASK A.12

   A. A only

   B. B only

   C. Both A and B

   D. Neither A nor B

Answer A is incorrect. Technician B is also correct.

Answer B is incorrect. Technician A is also correct.

Answer C is correct. Both Technicians are correct. The data link connector is a round 9-pin style on most late-model trucks. This connector is located somewhere inside the cab. Many computer systems allow the technician to use flash code diagnostics to retrieve trouble codes. Flash code diagnostics can be activated by signaling the computer.

Answer D is incorrect. Both Technicians are correct.

26. The function of a maxi-fuse is to:

TASK A.6

   A. Take the place of a circuit breaker.

   B. Close during a current overload.

   C. Open and close when signaled by a computer.

   D. Open an overloaded circuit when excessive current is present in the circuit.

Answer A is incorrect. Maxi-fuses are often used instead of fusible links. Maxi-fuses are not used in place of circuit breakers.

Answer B is incorrect. Maxi-fuses have a small metallic strip that burns up when high current is present in the circuit.

Answer C is incorrect. Maxi-fuses do not open and close while in operation. Maxi-fuses have continuity in normal operation, and they have a metallic strip that burns up when high current is present in the circuit.

Answer D is correct. Maxi-fuses have a metallic strip that burns up when the current flow in the circuit rises above the maxi-fuse rating. Once the fuse opens, it must be replaced to have continued electrical operation.

27. All of the following conditions could cause an undercharged battery EXCEPT:

TASK C.2

   A. Loose alternator bracket fasteners

   B. An oversized charging wire

   C. Worn alternator brushes

   D. Excessive voltage drop in the charging wire

Answer A is incorrect. A loose alternator bracket could cause an unwanted voltage drop in the charging ground circuit.

Answer B is correct. Using a larger charging wire would not cause a charging problem. A larger wire would carry the current more easily than a smaller wire.

Answer C is incorrect. Worn alternator brushes could reduce the charging output of the generator.

Answer D is incorrect. Excessive voltage drop in the charging wire would reduce the charge rate of the battery.

TASK C.3, C.7

28. An alternator with a 90 ampere rating produces 45 amps during an output test. A V-belt drives the alternator, and the belt is at the specified tension. Technician A says the V-belt may be worn and bottomed in the pulley. Technician B says the alternator pulley may be misaligned with the crankshaft pulley. Who is correct?

    A. A only

    B. B only

    C. Both A and B

    D. Neither A nor B

    Answer A is correct. Only Technician A is correct. Even though the belt is at the specified tension, it may slip because it is bottomed in the pulley. If a belt tension gauge is not available, belt tension can be determined by depressing the belt at the center of its span. As a rule, 3/8 inch (0.1 cm) is the approximate distance that the belt should be allowed to move.

    Answer B is incorrect. A misaligned pulley should not cause low alternator output because of belt slippage. It may, however, cause the belt to jump off (in which case there would be zero output) or it may cause rapid belt wear.

    Answer C is incorrect. Only Technician A is correct.

    Answer D is incorrect. Technician A is correct.

TASK C.4

29. When performing an alternator maximum output test, what is the most practical and safe way to make the alternator produce maximum output?

    A. Place a carbon pile tester across the battery terminals.

    B. Full-field the alternator.

    C. Turn on all of the vehicle's electrical loads.

    D. Install a discharged battery into the vehicle.

    Answer A is correct. A carbon pile tester can safely and quickly load the system to produce maximum output. The engine needs to be running at 1500–1600 rpm while performing this test.

    Answer B is incorrect. Even though full-fielding (where applicable) is an easy way to make the alternator produce maximum output, it is a potentially dangerous test: If the current is not dissipated somewhere, voltage in the system can rise to dangerous levels.

    Answer C is incorrect. All of the vehicle loads combined should not equal or exceed the alternator's maximum output.

    Answer D is incorrect. Temporarily installing a low battery will not necessarily force the alternator to maximum output. This would depend on the condition of the battery and how low its charge is.

30. Which of the following repair procedures would most likely correct an excessive voltage drop problem in the positive charging circuit?

TASK C.5

    A. Tighten the starter solenoid attaching bolts.

    B. Replace the terminal at the alternator output wire.

    C. Tighten the alternator mounting bracket.

    D. Replace the voltage regulator.

    Answer A is incorrect. Tightening the starter solenoid attaching bolts would not change the alternator output. Loose bolts in this area could cause the starter to fail to operate correctly.

    Answer B is correct. A faulty terminal on the alternator output wire could cause an excessive voltage drop on the positive charging circuit.

    Answer C is incorrect. Loose alternator mounting brackets could cause excessive voltage drop in the charging ground circuit, but not in the positive charging circuit.

    Answer D is incorrect. A defective voltage regulator would not cause an excessive voltage drop in the positive charging circuit. A defective voltage regulator could cause a no charge or an overcharge condition.

31. A medium-duty truck with a dead battery is being jump-started. Technician A says the engine should be running on the boost vehicle before attempting to crank the dead vehicle. Technician B says the engine should be off while connecting the booster cables. Who is correct?

TASK B.6

    A. A only

    B. B only

    C. Both A and B

    D. Neither A nor B

    Answer A is incorrect. Technician B is also correct.

    Answer B is incorrect. Technician A is also correct.

    Answer C is correct. Both Technicians are correct. Having the engine running on the boost vehicle is advisable because the charging system can charge the battery being boosted. It is a good practice to allow the boost vehicle to run 8–10 minutes on fast idle before attempting to start the dead vehicle. It is also a good practice to have the engine off while connecting the cables.

    Answer D is incorrect. Both Technicians are correct.

32. All of the following methods of wire repair in the charging system are currently used EXCEPT:

TASK C.7

    A. Crimp-and-seal connectors

    B. Butt connectors and tape

    C. New wiring harness

    D. Solder and heat shrink

    Answer A is incorrect. Crimp-and-seal connectors are acceptable to use in charging system repairs. These connectors are sealed against water intrusion.

    Answer B is correct. Butt connectors and tape are not used in professional wire repair practice because they will not prevent water from entering the connection.

    Answer C is incorrect. A new wiring harness is sometimes necessary to satisfy some manufacturers' requirements for charging system repairs.

    Answer D is incorrect. Solder and heat-shrink methods of wire repair are acceptable for charging system repairs. This repair technique does not allow water intrusion.

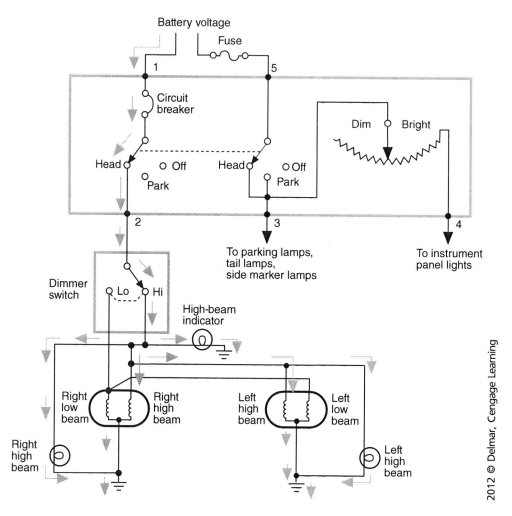

Battery voltage

To parking lamps, tail lamps, side marker lamps

To instrument panel lights

TASK D.1

33. The left-side headlight is dim only on the high beam in the figure above. The other headlights operate normally. Technician A says there may be high resistance in the left-side headlight ground. Technician B says there may be high resistance in the dimmer switch high-beam contacts. Who is correct?

A. A only

B. B only

C. Both A and B

D. Neither A nor B

Answer A is incorrect. A bad ground in the left headlight would affect the operation of both high beams and low beams.

Answer B is incorrect. High resistance in the dimmer switch high-beam contacts would affect both right- and left-side bulbs.

Answer C is incorrect. Neither Technician is correct.

Answer D is correct. Neither Technician is correct. The problem could be in the wiring that is common only to the left-side high-beam filaments or possibly with faulty bulbs. The other answers given would affect more than just the left high beams.

34. A truck has an intermittent fault with its high beams only. All of these could be a possible cause EXCEPT:

TASK D.3

A. A defective headlight dimmer switch

B. Defective high-beam filaments inside headlamps

C. A loose wiring harness connector

D. A defective headlight switch

Answer A is incorrect. A defective dimmer switch can cause intermittent high-beam operation.

Answer B is incorrect. A loose or otherwise faulty high-beam filament will cause partial or total failure of the high-beam circuit.

Answer C is incorrect. A loose connector in the high-beam circuit can affect its operation.

Answer D is correct. A defective headlight switch should affect both low- and high-beam operation equally, because it powers both circuits from the same contacts.

35. A trailer has inoperative taillights on one side only. Technician A says to check the trailer circuit connector for an open. Technician B uses an ohmmeter to check the continuity between the defective side and the trailer circuit connector with the circuit under power. Who is correct?

TASK D.4

A. A only

B. B only

C. Both A and B

D. Neither A nor B

Answer A is incorrect. An open in the trailer connector would cause both taillights to be inoperative.

Answer B is incorrect. An ohmmeter cannot be used to check resistance of a circuit under power.

Answer C is incorrect. Neither Technician is correct.

Answer D is correct. Neither Technician is correct. This problem could be caused by a faulty taillight bulb, taillight socket, or a wiring problem involving only the side affected.

36. A truck electrical system is being repaired. Technician A says that wiring schematics give exact details of the location of all electrical components on a truck. Technician B says that a wiring schematic will usually contain pin-out test procedures that can be used to troubleshoot many electrical faults. Who is correct?

TASK A.9

A. A only

B. B only

C. Both A and B

D. Neither A nor B

Answer A is incorrect. A wiring schematic is a drawing that shows the wires and components in a circuit using symbols to represent the components.

Answer B is incorrect. A wiring schematic does not typically contain a pin-out procedure.

Answer C is incorrect. Neither Technician is correct.

Answer D is correct. Neither Technician is correct. A wiring schematic is a drawing that shows the wires and components in a circuit using symbols to represent the components. These schematics are very useful in the diagnosis of electrical problems on trucks because a technician can use them to decide where to begin his/her testing steps.

TASK D.8

37. If only the right side of the trailer's turn signals is illuminated, the technician should:

    A. Replace the turn signal flasher.
    B. Replace the bulbs on the left side of the trailer.
    C. Check the trailer electrical connection.
    D. Check the brake light switch for proper operation.

    Answer A is incorrect. A defective turn signal flasher will affect the trailer lights on both sides, not just one.

    Answer B is incorrect. It is not likely that all the bulbs on the left side have burned out. They need to be tested to be sure that they are getting proper voltage before being replaced.

    Answer C is correct. The trailer's electrical connector is one area where the two different circuits are independent of one other. If the pin for the left turn signal circuit has corrosion or is broken, this would be a good reason for the lights on the left side to be inoperative.

    Answer D is incorrect. A faulty brake light switch should affect both turn signal circuits in the same manner.

TASK D.11

38. A trailer light cord power connector needs to be tested. Technician A says a digital multimeter (DMM) is an effective tool for testing the cord. Technician B says an incandescent test light is an effective method for testing the cord. Who is correct?

    A. A only
    B. B only
    C. Both A and B
    D. Neither A nor B

    Answer A is incorrect. A DMM is a good test instrument; however, the most common failure in the trailer light cord is reduced current flow due to corrosion. A DMM will not require sufficient current flow to test this circuit accurately.

    Answer B is correct. Technician B is correct. A test light is the best method to test the trailer light cord. The test light will provide sufficient current demand to test the circuit.

    Answer C is incorrect. Only Technician B is correct.

    Answer D is incorrect. Technician B is correct.

TASK E.1

39. A heavy truck is being diagnosed for a fuel gauge that does not work. Technician A says that the other gauges should be checked for correct operation as part of the diagnosis process. Technician B says that the sending unit for the fuel gauge is in the fuel tank. Who is correct?

    A. A only
    B. B only
    C. Both A and B
    D. Neither A nor B

    Answer A is incorrect. Technician B is also correct.

    Answer B is incorrect. Technician A is also correct.

    Answer C is correct. Both Technicians are correct. It is a good diagnostic technique to verify that the other gauges are functioning correctly. The sending unit for the fuel gauge is located in the fuel tank.

    Answer D is incorrect. Both Technicians are correct.

40. On a truck with electronic instrumentation, gauge accuracy is suspect. What would be the recommended method to determine where the fault lies?

TASK E.2

   A. Swap the panel with a new one.

   B. Replace the suspect sender unit(s).

   C. Ground the sender wire at the suspect sender unit(s).

   D. Connect a scan tool and compare display to gauge readings.

   Answer A is incorrect. While swapping panels may help to locate a defective gauge, it is not good practice because electrical/electronic components are not returnable.

   Answer B is incorrect. As with the previous answer, it is bad practice to troubleshoot a problem by trial and error.

   Answer C is incorrect. With many senders on an electronic engine, grounding is not possible.

   Answer D is correct. By using the scan tool, the technician can troubleshoot the problem using the OEM recommended procedure. If the diagnostic tool shows normal readings, then the gauge would be suspect.

41. A truck with a data bus network problem is being diagnosed. Technician A says that the driver will not likely notice any unusual problems if the data bus wires become shorted together. Technician B says data bus communication happens when voltage pulses are sent from module to module thousands of times per second. Who is correct?

TASK A.10

   A. A only

   B. B only

   C. Both A and B

   D. Neither A nor B

   Answer A is incorrect. The driver would likely notice problems in several systems, including the engine, transmission, instrument cluster, and climate control.

   Answer B is correct. Only Technician B is correct. Data bus communication takes place over one or two wires by pulsing voltage at a very fast rate. Some networks can communicate up to 250,000 times per second. This rate is getting faster as trucks become more and more complicated and rely more on computer-controlled component modules.

   Answer C is incorrect. Only Technician B is correct.

   Answer D is incorrect. Technician B is correct.

42. The "check engine" light (CEL) illuminates while a truck is being operated. Which of the following causes would not require the driver's immediate attention?

TASK E.4

   A. Low engine oil pressure

   B. High engine coolant temperature

   C. Maintenance reminder

   D. Low coolant level

   Answer A is incorrect. Low engine oil pressure would require immediate action from the driver.

   Answer B is incorrect. High engine water temperature would require an immediate response from the driver.

   Answer C is correct. A maintenance reminder is one example of a CEL illumination that does not require the driver's immediate attention.

   Answer D is incorrect. Low coolant level is a potentially damaging engine condition that requires immediate attention.

TASK E.7

43. A vehicle electric horn does not function when the horn is depressed. Technician A uses a test lamp to check the power and ground terminals of the horn relay. Technician B uses a digital multimeter (DMM) to check the power and ground terminals of the horn relay. Who is correct?

   A. A only
   B. B only
   C. Both A and B
   D. Neither A nor B

Answer A is incorrect. Technician B is also correct.

Answer B is incorrect. Technician A is also correct.

Answer C is correct. Both Technicians are correct. A test lamp is a good way to test for proper power and grounds at the relay. A DMM is another good way to test for power and grounds at the relay.

Answer D is incorrect. Both Technicians are correct.

TASK E.8

44. An older-type, two-speed wiper motor is being diagnosed. Technician A says that some trucks create two-speed operation by using separate sets of high- and low-speed brushes inside the motor. Technician B states that the two-speed operation is accomplished on some trucks by using an external resistor pack similar to a heater blower motor. Who is correct?

   A. A only
   B. B only
   C. Both A and B
   D. Neither A nor B

Answer A is incorrect. Technician B is also correct.

Answer B is incorrect. Technician A is also correct.

Answer C is correct. Both Technicians are correct. There are two ways to control wiper speed on older trucks: Use high- and low-speed brushes, or use an external resistor pack. Most manufacturers now use the high- and low-speed brush design on late-model trucks.

Answer D is incorrect. Both Technicians are correct.

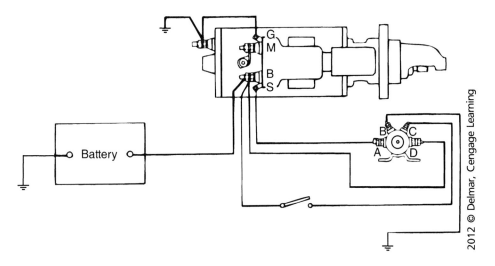

2012 © Delmar, Cengage Learning

45. A technician is about to perform a voltage drop test across the starter solenoid internal contacts. Referring to the figure above, where should the voltmeter leads be placed?

TASK B.12

   A. Between the positive battery terminal and point B

   B. Between the positive battery terminal and point M

   C. Between points B and M

   D. Between points G and ground

Answer A is incorrect. Placing the voltmeter leads between the positive battery terminal and point B would test the voltage drop in the positive battery cable.

Answer B is incorrect. Placing the voltmeter leads between the positive battery terminal and point M would test the combined voltage drop of both the positive battery cable and the solenoid internal contacts.

Answer C is correct. By probing between points B and M, the voltage drop across the solenoid internal contacts can be tested with the starter cranking.

Answer D is incorrect. Probing between points G and ground will only test the voltage drop in the solenoid ground wire.

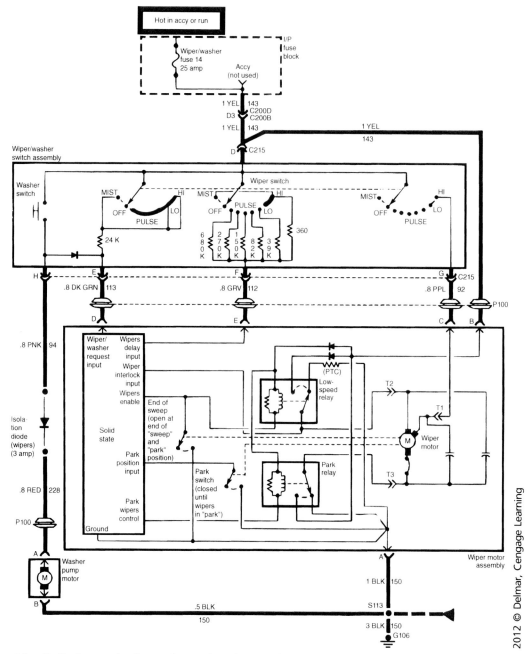

TASK E.10

46. Referring to the figure above, the wiper washer pump motor runs constantly. Which of the following conditions would most likely cause this problem?

A. The ground side of the motor is shorted to ground.

B. The control switch is shorted to ground.

C. The contacts in the switch are stuck closed.

D. The wiper washer pump relay contacts are stuck closed.

Answer A is incorrect. If the ground side of the motor shorted to ground, there would be no adverse reaction. This system does not use ground-side switching.

Answer B is incorrect. If the control switch shorted to ground, the fuse would blow.

Answer C is correct. Stuck closed switch contacts would cause a continuous circuit and constant pump operation.

Answer D is incorrect. There is no separate relay in this system for the wiper washer pump motor.

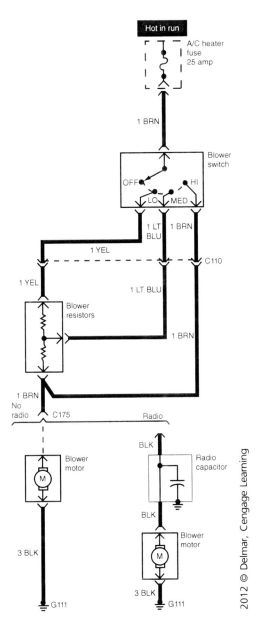

47. Referring to the figure above, when the blower resistors are removed, the blower motor will:

   A. Not operate at all.

   B. Blow the system fuse.

   C. Operate on high speed only.

   D. Operate on low speed only.

TASK E.12

Answer A is incorrect. The blower motor will operate on the high-speed position because the resistors are by-passed.

Answer B is incorrect. Removing the resistors will create an open circuit in the low and medium speeds. Since there would be no current flow, the fuse would not blow.

Answer C is correct. With the switch in the high-speed position, the resistors are by-passed and the motor will operate normally, but only at high speed.

Answer D is incorrect. With the resistors removed, current cannot flow from the low-speed switch position to the motor.

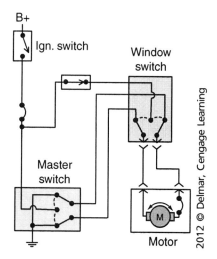

TASK E.14

48. Referring to the figure above, the power window will not operate from either the master or window switches. Which of the following would be the most likely cause of this problem?

A. Open diode
B. Bad ground at the master switch
C. Closed circuit breaker at the motor
D. Shorted diode

Answer A is incorrect. An open diode would have no effect on the power window operation when using the master switch.

Answer B is correct. A bad ground at the master switch would cause the power window motor to be totally inoperative. The ground at the master switch is the only ground provided to the power window circuit.

Answer C is incorrect. A closed circuit breaker would not cause the power window motor to be inoperative.

Answer D is incorrect. A shorted diode would have no effect on the power window operation when using the master switch.

TASK A.2

49. Technician A says an ammeter is used to test continuity. Technician B says an ammeter measures current flow in a circuit. Who is correct?

A. A only
B. B only
C. Both A and B
D. Neither A nor B

Answer A is incorrect. An ammeter is used to measure the current flow in an electrical circuit. The ammeter would not be a useful tool to use when testing for continuity.

Answer B is correct. Only Technician B is correct. The ammeter is used to measure the electrical current flow in a circuit. Ammeters are typically connected in a series with the circuit on lower amperage circuits. Amp clamps can be used on circuits that draw more current.

Answer C is incorrect. Only Technician B is correct.

Answer D is incorrect. Technician B is correct.

50. Where would the power door lock relay most likely be located on the truck?

TASK E.18

  A. Under the driver's seat
  B. In the under-hood power distribution center
  C. Near the rear heating, ventilating, and air conditioning (HVAC) assembly
  D. In the in-cab power distribution center

Answer A is incorrect. The power door lock relay would not be commonly located under the driver's seat.

Answer B is incorrect. The power door lock relay would not be commonly located in the under-hood power distribution center.

Answer C is incorrect. The power door lock relay would not be located near the rear heating, ventilating, and air conditioning (HVAC) assembly.

Answer D is correct. The power door lock relay would likely be located inside the cab area near the other electrical components. This relay is often located in the in-cab power distribution center.

PREPARATION EXAM 4—ANSWER KEY

1.	D	**18.**	D	**35.**	A
2.	B	**19.**	C	**36.**	D
3.	B	**20.**	D	**37.**	C
4.	A	**21.**	A	**38.**	B
5.	C	**22.**	B	**39.**	B
6.	B	**23.**	A	**40.**	A
7.	B	**24.**	D	**41.**	A
8.	A	**25.**	A	**42.**	A
9.	C	**26.**	D	**43.**	B
10.	D	**27.**	C	**44.**	C
11.	C	**28.**	A	**45.**	B
12.	A	**29.**	C	**46.**	C
13.	A	**30.**	C	**47.**	C
14.	A	**31.**	B	**48.**	B
15.	D	**32.**	A	**49.**	C
16.	D	**33.**	B	**50.**	B
17.	D	**34.**	C		

PREPARATION EXAM 4—EXPLANATIONS

TASK A.1, A.3

1. Technician A says that a self-powered test lamp can be used to check continuity in a circuit managed by an electronic control module (ECM). Technician B states that an analog multimeter may be used to test voltages in an electronic circuit. Who is correct?

   A.   A only

   B.   B only

   C.   Both A and B

   D.   Neither A nor B

   Answer A is incorrect. A self-powered test light should never be used on a circuit that involves an ECM because damage to the electronic circuits will occur.

   Answer B is incorrect. An analog multimeter should never be used on a circuit that is controlled by an ECM. Analog meters are not typically high-impedance tools and thus will cause excessive current flow in electronic circuits.

   Answer C is incorrect. Neither Technician is correct.

   Answer D is correct. Neither Technician is correct. Only high-impedance digital multimeters should be used on circuits that are controlled by an ECM.

2. All of the following could be used when checking a circuit for continuity EXCEPT:

TASK A.3

   A. An ohmmeter

   B. An ammeter

   C. A voltmeter

   D. A self-powered test light

   Answer A is incorrect. An ohmmeter can be used to test for continuity of switches and wires. The leads should be placed in parallel with the item to be tested; a test result of low ohms will prove that the item has continuity.

   Answer B is correct. An ammeter is not used to test for continuity. Ammeters are used to test for current flow in an electrical circuit.

   Answer C is incorrect. A voltmeter can be used to test for continuity by checking the voltage drops in the circuit.

   Answer D is incorrect. A self-powered test light could be used to test for continuity by connecting the tester to each end of the switch or wire. A circuit that has continuity will cause the light to illuminate. A self-powered test light should never be used on a powered circuit.

3. A truck's electronic cruise control will not operate, although everything else functions properly. Which of the following should a technician check first?

TASK E.19

   A. Servo motor

   B. Diagnostic software

   C. Linkage to the fuel pump

   D. Vehicle speed sensor

   Answer A is incorrect. An electronic cruise control does not use a servo motor. Cruise control is managed by the ECM.

   Answer B is correct. The first step in troubleshooting electronically controlled engines is to connect the vehicle data link and access the diagnostic software.

   Answer C is incorrect. An electronically controlled engine has no mechanical linkage to the fuel pump.

   Answer D is incorrect. A defective vehicle speed sensor would affect other chassis systems and log a fault code.

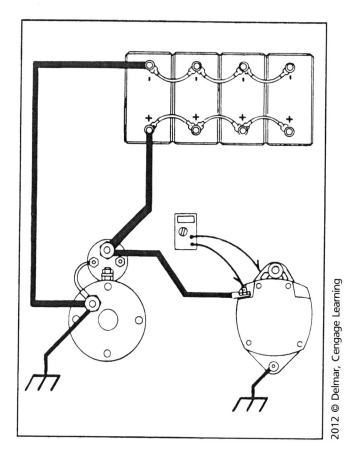

2012 © Delmar, Cengage Learning

TASK A.1

4. Referring to the figure above, what test is being performed?

A. Voltage output test

B. Positive charging circuit cable voltage drop test

C. Charging ground circuit voltage drop test

D. Starter operating voltage test

Answer A is correct. The charging system voltage is being measured in the figure because one lead is connected to the output terminal and the other lead is connected to ground.

Answer B is incorrect. To perform a positive charging circuit cable voltage drop test, one voltmeter probe would be at the "battery" terminal on the starter solenoid and the other probe would be at the output terminal of the alternator.

Answer C is incorrect. To measure the voltage drop of the charging ground circuit, you would probe at the alternator housing with one lead and at the battery negative terminal with the other lead.

Answer D is incorrect. To measure starter operating voltage, you would probe at the starter, not the alternator.

5. Technician A says that a poor ground connection will cause reduced current flow in an electrical circuit. Technician B says that water corrosion in the wiring will cause reduced current flow in an electrical circuit. Who is correct?

TASK A.2

   A. A only
   B. B only
   C. Both A and B
   D. Neither A nor B

   Answer A is incorrect. Technician B is also correct.

   Answer B is incorrect. Technician A is also correct.

   Answer C is correct. Both Technicians are correct. A poor ground and corroded wiring will both cause decreased current flow in an electrical circuit due to the increased resistance associated with both of these problems.

   Answer D is incorrect. Both Technicians are correct.

6. A resistance test was performed on an open headlight switch and the result is 5.857 megohms. Technician A says that the switch is faulty because the reading is beyond the specifications. Technician B says that the switch has over 5 million ohms of resistance. Who is correct?

TASK A.3

   A. A only
   B. B only
   C. Both A and B
   D. Neither A nor B

   Answer A is incorrect. An open switch is supposed to have very high resistance.

   Answer B is correct. Only Technician B is correct. The reading is over 5 million ohms. An open switch should have very high resistance.

   Answer C is incorrect. Only Technician B is correct.

   Answer D is incorrect. Technician B is correct.

7. A truck is being diagnosed for a potential volt gauge problem. Technician A says that it is normal for the dash voltmeter to read a different voltage from that of a test voltmeter across the battery terminals. Technician B says that resistance in the dash voltmeter ground will affect its readings. Who is correct?

TASK C.1

   A. A only
   B. B only
   C. Both A and B
   D. Neither A nor B

   Answer A is incorrect. The dash voltmeter reads battery voltage.

   Answer B is correct. Only Technician B is correct. Unwanted resistance in the voltmeter circuit will cause a voltage drop, which will affect the gauge reading. The voltmeter alone should not be used to indicate a charging system problem. A volt reading of 14 volts might indicate a good charging system; however, the AC generator might be putting out only enough current to maintain a proper voltage level if there are no loads, thus leading you to a wrong conclusion.

   Answer C is incorrect. Only Technician B is correct.

   Answer D is incorrect. Technician B is correct.

TASK A.4

8. Technician A says that a short to ground before the load will cause the circuit protection device to open when the hot-side switch is turned on. Technician B says that a short to ground after the load will cause the circuit protection device to open when the hot-side switch is turned on. Who is correct?

    A. A only
    B. B only
    C. Both A and B
    D. Neither A nor B

    Answer A is correct. Only Technician A is correct. This scenario would cause excessive current to flow because the electricity would take the path of least resistance and by-pass the load. When this happens, the circuit protection device would heat up and open the circuit.

    Answer B is incorrect. A grounded circuit after the load would not negatively affect the circuit. This would just be a redundant ground located in the ground side of the circuit.

    Answer C is incorrect. Only Technician A is correct.

    Answer D is incorrect. Technician A is correct.

TASK B.15

9. Technician A says that a mistimed engine can cause slow cranking speed. Technician B says that low engine compression can cause rapid cranking speed. Who is correct?

    A. A only
    B. B only
    C. Both A and B
    D. Neither A nor B

    Answer A is incorrect. Technician B is also correct.

    Answer B is incorrect. Technician A is also correct.

    Answer C is correct. Both Technicians are correct. An engine that is mistimed will sometimes crank slower than normal due to the ignition system firing at the wrong time. An engine with low compression will often have rapid cranking speed due to less energy required to move the pistons to the top of the compression stroke.

    Answer D is incorrect. Both Technicians are correct.

TASK A.6

10. Technician A says that any kind of wire can be used to replace a defective fusible link as long as the gauge is one size smaller than the circuit being protected. Technician B says that circuit breakers should always be replaced after repairing a short to ground problem. Who is correct?

    A. A only
    B. B only
    C. Both A and B
    D. Neither A nor B

    Answer A is incorrect. Only the correct size fusible link should be used when replacing a defective fusible link.

    Answer B is incorrect. It is not always necessary to replace the circuit breaker after repairing a short to ground problem. Some circuit breakers have to be manually reset after opening due to high current.

    Answer C is incorrect. Neither Technician is correct.

    Answer D is correct. Neither Technician is correct. The correct size fusible link should be used when making repairs to fusible link circuits. It is not necessary to replace circuit breakers after repairing a short to ground problem. If the circuit breaker is a manual reset design, then the circuit breaker can be reset to continue operation. Automatic circuit breakers may need to be replaced after a short to ground problem is repaired due to the rapid operation of these devices when a short to ground exists. The rapid operation can damage the contacts.

11. A truck has a turn signal complaint. The left-front light does not work and the left-rear light flashes slower than normal. Technician A says the left-front bulb could be defective. Technician B says there could be an open circuit between the switch and the left-front bulb. Who is correct?

TASK D.9

  A. A only
  B. B only
  C. Both A and B
  D. Neither A nor B

Answer A is incorrect. Technician B is also correct.

Answer B is incorrect. Technician A is also correct.

Answer C is correct. Both Technicians are correct. If the front bulb fails, this will reduce current demand in the circuit, causing the flasher to blink slower than normal. An open circuit to the front bulb will reduce the current demand in the circuit, making the flasher blink slower than normal and also not allow the front bulb to light.

Answer D is incorrect. Both Technicians are correct.

12. Which of the following would be the LEAST LIKELY test result of a good International Organization for Standardization (ISO) relay?

TASK A.8

  A. 1 ohm when connected to terminals 30 and 85
  B. Out of limits (OL) when connected to terminals 30 and 87
  C. 80 ohms when connected to terminals 85 and 86
  D. 1 ohm when connected to terminals 30 and 87a

Answer A is correct. The ohmmeter reading for terminals 30 and 85 should be OL. Terminal 30 is on the "load" side of the relay and terminal 85 is on the "coil" side of the relay.

Answer B is incorrect. Terminals 30 and 87 are the normally open (NO) contacts of the relay. The resistance at these two terminals should be OL.

Answer C is incorrect. Terminals 85 and 86 are the coil connections of the relay. Most relay coils have approximately 60 to 90 ohms of resistance.

Answer D is incorrect. Terminals 30 and 87a are the normally closed (NC) contacts of the relay, which should have very low resistance during this test.

13. The temperature light stays on continuously while driving a heavy-duty truck. Technician A says that a shorted wire leading to the temperature sending unit could be the cause. Technician B says that an "open" temperature sending unit could be the cause. Who is correct?

TASK E.4

  A. A only
  B. B only
  C. Both A and B
  D. Neither A nor B

Answer A is correct. Only Technician A is correct. A shorted wire could cause the temperature light to stay on continuously. The temperature sending unit causes the light to illuminate by providing a ground for the bulb.

Answer B is incorrect. An "open" temperature sending unit would cause the light to not come on at all.

Answer C is incorrect. Only Technician A is correct.

Answer D is incorrect. Technician A is correct.

TASK A.10

14. A truck is being diagnosed for a "no communication" fault on the scan tool. Technician A says that an open terminating resistor could be the cause. Technician B says that a secure terminal connection at the splice block could be the cause. Who is correct?

    A. A only
    B. B only
    C. C. Both A and B
    D. Neither A nor B

    Answer A is correct. Only Technician A is correct. An open terminating resistor could potentially cause communication problems on the data bus network. The J1939 data bus should have approximately **60** ohms of resistance when tested with an ohmmeter. This test can be performed at the data connector and should be done only after disconnecting the truck batteries.

    Answer B is incorrect. A tight terminal connection at a splice block would not cause a communication problem for the data bus network. Secure terminal connections are a desirable quality for all electrical networks.

    Answer C is incorrect. Only Technician A is correct.

    Answer D is incorrect. Technician A is correct.

TASK B.2

15. A technician is preparing to load test the batteries on a truck with four batteries. Technician A says that trucks with multiple batteries should be tested with the batteries connected together. Technician B says that it is normal to see sparks while the load test is being performed. Who is correct?

    A. A only
    B. B only
    C. Both A and B
    D. Neither A nor B

    Answer A is incorrect. Most battery testers will require the technician to separate the batteries before performing a load test.

    Answer B is incorrect. Sparks present in an electrical circuit during testing (or any time) indicate a bad connection.

    Answer C is incorrect. Neither Technician is correct.

    Answer D is correct. Neither Technician is correct. Trucks with multiple batteries need to have the batteries separated before extensive battery testing happens. If sparks are present during a load test, then the test needs to be aborted and the connections checked closely.

TASK B.1

16. Technician A says that measuring the terminal post voltage while the battery is being charged is a good way to determine the battery state of charge. Technician B says that terminal post voltage often drops down to approximately 7.5 volts while the engine is cranking. Who is correct?

    A. A only
    B. B only
    C. Both A and B
    D. Neither A nor B

    Answer A is incorrect. Measuring the terminal post voltage while the battery is being charged will not reveal the state of charge. The voltage must be checked before the charger is connected to determine the state of charge.

    Answer B is incorrect. The terminal post voltage should not drop to 7.5 volts while cranking the engine.

    Answer C is incorrect. Neither Technician is correct.

    Answer D is correct. Neither Technician is correct. The battery state of charge can be tested by checking the post voltage prior to charging the battery. The battery voltage should not drop below approximately 10.5 volts on a sufficiently charged battery.

17. The alternator needs to be replaced on a truck. Technician A says that the battery cables do not have to be disconnected while performing the exchange. Technician B says that the pulley diameter does not have to match when installing the new alternator. Who is correct?

TASK C.6

   A. A only

   B. B only

   C. Both A and B

   D. Neither A nor B

Answer A is incorrect. It is dangerous to replace an alternator without disconnecting the battery cables.

Answer B is incorrect. The alternator pulley on the replacement device should be the same diameter as the old alternator.

Answer C is incorrect. Neither Technician is correct.

Answer D is correct. Neither Technician is correct. It is recommended to disconnect the negative battery cable before beginning the replacement of an alternator. This prevents accidentally grounding out the charge wire during the process. The replacement alternator should have the same diameter pulley as the old alternator to assure the correct charge and performance.

18. Where is the low voltage disconnect (LVD) module connected to the truck?

   A. In series between the alternator and the battery pack

   B. In parallel with the battery pack and the truck power distribution center

   C. In parallel with the alternator and the battery pack

TASK B.7

   D. In series between the battery pack and the truck power distribution center

Answer A is incorrect. The LVD module could not shut off power to the truck if it was located between the alternator and the battery pack.

Answer B is incorrect. The module is connected in series, not in parallel.

Answer C is incorrect. The module must be connected in series in a location that would make it possible to remove power from the truck accessories.

Answer D is correct. The LVD module is mounted in series between the battery and the power distribution center. When battery voltage falls below a certain level, the module opens the connection and shuts off battery power to the truck accessories.

19. A truck is being diagnosed for a slow crank problem. Technician A says that testing the voltage drop on the battery cables while cranking the engine will reveal cable problems. Technician B says that worn starter brushes could cause the slow crank problem. Who is correct?

TASK B.9,
B.15

   A. A only

   B. B only

   C. Both A and B

   D. Neither A nor B

Answer A is incorrect. Technician B is also correct.

Answer B is incorrect. Technician A is also correct.

Answer C is correct. Both Technicians are correct. Performing voltage drop tests on battery cables while cranking the engine is an accurate method of testing the battery cables. There should not be more than 0.5 volts on either cable during this test. Worn starter brushes could cause a slow crank problem by lowering the starter current draw due to the increased electrical resistance.

Answer D is incorrect. Both Technicians are correct.

TASK B.10,
B.15

20. A truck will not crank and the technician notices that the interior lights do not dim when the ignition switch is moved to the start position. Of the following, what is the most likely cause of this condition?

A. Stuck closed starter relay
B. Loose battery cable connections
C. Loose starter mounting bolts
D. Stuck open start switch

Answer A is incorrect. A stuck closed starter relay would cause the starter to hang in the engaged position and would cause the starter motor to burn up.

Answer B is incorrect. Loose battery cable connections could cause a no crank problem, but the dome lights would get dimmer or go out when the ignition switch was operated.

Answer C is incorrect. Loose starter mounting bolts could cause unusual starter sounds, but would not typically cause a no crank problem.

Answer D is correct. A start switch that is stuck open would cause this problem. No signal would be sent to the starter relay or solenoid, therefore the interior lights would not change as the start switch is operated.

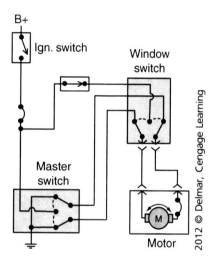

21. Referring to the figure above, the power window operates normally from the master switch, but does not operate using the window switch. Which of the following could be the cause?

TASK E.14,
E.15

A. An open between the ignition switch and window switch
B. An open in the window switch movable contacts
C. An open in the master switch ground wire
D. A short to ground at the circuit breaker in the motor

Answer A is correct. An open between the ignition switch and the window switch will cause only the window switch to not operate the motor.

Answer B is incorrect. An open in the window switch movable contacts will open the master switch to the motor circuit.

Answer C is incorrect. An open in the master switch ground wire will result in no motor operation when the master switch is depressed.

Answer D is incorrect. A short to ground at the circuit breaker will result in a blown fuse and no motor operation at all.

22. Which of the following would be the recommended method to use when professionally cleaning battery cable clamps?

    A. Pneumatic die grinder
    B. Wire brush
    C. Carburetor cleaner and scraper
    D. Screwdriver

TASK B.13

Answer A is incorrect. A power tool should never be used around a battery due to the danger of possibly producing sparks, as well as causing flying debris in this area.

Answer B is correct. A wire brush can safely be used to remove corrosion and residue from the battery terminals and cable ends.

Answer C is incorrect. Carburetor cleaner should never be used near a battery due to the chemical properties, which could react to the chemicals associated with the battery.

Answer D is incorrect. A screwdriver would not be an effective tool to use to clean the battery cable clamps.

23. All of the following parameters can be calibrated with a scan tool on some late-model trucks EXCEPT:

    A. Engine coolant temperature
    B. Maximum cruise control speed
    C. Maximum vehicle speed
    D. Maximum idle time

TASK A.12

Answer A is correct. The engine coolant temperature is not a typical parameter that can be calibrated with a scan tool.

Answer B is incorrect. The maximum cruise control speed is a value that can be calibrated with a scan tool.

Answer C is incorrect. The maximum vehicle speed is a value that can be calibrated with a scan tool.

Answer D is incorrect. The maximum idle time is a value that can be calibrated with a scan tool.

24. Technician A says that a broken wire to a door ajar switch will likely cause a parasitic draw problem. Technician B says that a blown horn fuse will likely cause a parasitic draw problem. Who is correct?

    A. A only
    B. B only
    C. Both A and B
    D. Neither A nor B

TASK A.5

Answer A is incorrect. A broken wire to a door ajar switch would not likely cause a parasitic draw problem. The dome light would not function with that door open.

Answer B is incorrect. A blown horn fuse would not likely cause a parasitic draw problem.

Answer C is incorrect. Neither Technician is correct.

Answer D is correct. Neither Technician is correct. A broken wire at a door ajar switch would cause the dome light to not function when that door is opened. A blown horn fuse would cause an inoperative horn. Neither of these problems would cause a parasitic draw problem.

TASK A.11,
A.12

25. A truck is being diagnosed for an antilock brake system (ABS) indicator that is staying illuminated. All of the following steps are advisable to perform in the early stages of this diagnostic procedure EXCEPT:

A. Use an ohmmeter to measure the resistance of all ABS speed sensors.
B. Connect a scan tool to the truck to retrieve the diagnostic trouble code (DTC).
C. Press the code activation switch to engage flash code diagnostics.
D. Perform a visual inspection of the ABS system to identify obvious defects.

Answer A is correct. It is not advisable to measure the resistance of sensors during the early stages of a diagnostic routine. This may be necessary after several of the preliminary steps are performed.

Answer B is incorrect. Using a scan tool to retrieve DTCs is a common step in the troubleshooting process of most computer-controlled systems.

Answer C is incorrect. Some computer-controlled systems display DTCs by using flash code diagnostics. This process is performed by depressing the code activation switch and watching the correct indicator blink a numbered sequence.

Answer D is incorrect. Performing a good visual inspection of a suspect ABS system is an advisable step to follow when troubleshooting. The technician should look for broken or damaged components, as well as faulty wiring.

TASK A.3

26. Technician A says that an electrical switch that has continuity will have reduced current flow. Technician B says that a piece of wire that has high resistance will have increased current flow. Who is correct?

A. A only
B. B only
C. Both A and B
D. Neither A nor B

Answer A is incorrect. A switch that has continuity will have normal current flow. A closed switch should have continuity.

Answer B is incorrect. A wire that has high resistance will have reduced current flow due to the increased resistance.

Answer C is incorrect. Neither Technician is correct.

Answer D is correct. Neither Technician is correct. A closed switch should have continuity, which will allow normal current to flow in a circuit. A wire with added electrical resistance will have reduced current flow.

TASK C.2

27. Technician A says that a poor connection at the charging output wire can cause a charging problem. Technician B says that many charging circuits contain a circuit protection device. Who is correct?

A. A only
B. B only
C. Both A and B
D. Neither A nor B

Answer A is incorrect. Technician B is also correct.

Answer B is incorrect. Technician A is also correct.

Answer C is correct. Both Technicians are correct. A poor connection at the charging output wire will cause an unwanted voltage drop, which will reduce the charging capacity to the battery. Most charging circuits have either a fusible link or a large fuse that protects that circuit in case of a grounding problem.

Answer D is incorrect. Both Technicians are correct.

28. The alternator belt is being tightened on a truck that uses V-belts. Technician A says that overtightening the alternator belt could cause bearing failure in the alternator. Technician B says that undertightening the alternator belt could cause bearing failure in the alternator. Who is correct?

TASK C.3

A. A only

B. B only

C. Both A and B

D. Neither A nor B

Answer A is correct. Only Technician A is correct. If an alternator belt is tightened beyond the recommended specification, then the alternator bearing will have to endure an excessive load, which would cause it to fail sooner than normal.

Answer B is incorrect. Undertightening an alternator will cause the belt to slip when in operation and loaded. This will be annoying, cause early belt failure, and likely cause an undercharged battery, but it will not cause bearing failure.

Answer C is incorrect. Only Technician A is correct.

Answer D is incorrect. Technician A is correct.

29. A full-fielded alternator could cause all of the following conditions EXCEPT:

TASK C.4

A. High battery voltage level

B. Battery gassing

C. Low battery voltage level

D. High alternator amperage output

Answer A is incorrect. A full-fielded alternator will produce maximum output regardless of voltage. This will cause the battery voltage to rise above safe levels.

Answer B is incorrect. A battery that is being overcharged can cause the battery to boil and gas.

Answer C is correct. A full-fielded alternator will cause battery voltage to rise to the maximum charge output.

Answer D is incorrect. A full-fielded alternator will cause high amperage output from the alternator.

30. A truck is being diagnosed for a charging problem. The alternator only charges at 11.8 volts. Technician A says that the charging output wire may have an excessive voltage drop. Technician B says that the charging ground circuit may have an excessive voltage drop. Who is correct?

TASK C.5

A. A only

B. B only

C. Both A and B

D. Neither A nor B

Answer A is incorrect. Technician B is also correct.

Answer B is incorrect. Technician A is also correct.

Answer C is correct. Both Technicians are correct. Excessive voltage drop in either the positive or ground side could cause an alternator output problem. The maximum voltage drop allowable is 0.5 volts in either side of the circuit while charging at a high rate.

Answer D is incorrect. Both Technicians are correct.

TASK B.3

31. Which of the following practices will reduce the chances of a battery cable connection getting corroded in the future?

    A. Using baking soda on the battery tray while the battery is removed
    B. Applying terminal protection spray after connecting the terminals to the batteries
    C. Applying bearing grease on the battery terminals prior to connecting the batteries
    D. Applying battery terminal cleaning spray after connecting the terminals to the batteries

    Answer A is incorrect. Baking soda will neutralize battery acid, which will assist in cleaning up the area around the battery.

    Answer B is correct. It is a good practice to apply terminal protection spray to the battery terminals after connecting the terminals to the batteries. This substance acts as a barrier to prevent corrosion from building up on the terminals.

    Answer C is incorrect. Bearing grease should not be used on the battery terminals prior to connecting the terminals. This practice would create unwanted resistance in the battery connection.

    Answer D is incorrect. Battery terminal cleaning spray is used to assist in cleaning the battery terminals during a battery service. It is generally used before the terminals are reconnected to the battery.

TASK C.7

32. A truck requires a new charging circuit connector at the alternator. Technician A says that the negative battery cable should be disconnected before beginning the repair. Technician B says that a butt loop connector is the recommended part for making the repair. Who is correct?

    A. A only
    B. B only
    C. Both A and B
    D. Neither A nor B

    Answer A is correct. Only Technician A is correct. The negative battery cable should be disconnected before beginning the repair of the charging circuit connector. This will keep the technician from grounding out the charge wire with the wrench by touching a metallic part of the engine.

    Answer B is incorrect. A butt connector that is not sealed against water intrusion is not recommended on under-the-hood repairs. A crimp-and-seal connector or heat shrink should be used to keep water out of the wiring.

    Answer C is incorrect. Only Technician A is correct.

    Answer D is incorrect. Technician A is correct.

TASK D.4

33. Which dual-filament bulb design has a round base with two contacts on the bottom and receives its ground through the case of the base?

    A. 1156
    B. 1157
    C. 3056
    D. 3057

    Answer A is incorrect. An 1156 bulb has a round base, but only has one contact and one filament.

    Answer B is correct. This dual-filament bulb is used in many applications on trucks and trailers. The round base has staggered indexing pins to assure that it is installed correctly.

    Answer C is incorrect. A 3056 bulb has a rectangular base and is a single-filament bulb.

    Answer D is incorrect. A 3057 bulb has a rectangular base and is a dual-filament bulb.

34. Technician A says that the dash light dimmer is sometimes built into the headlight switch. Technician B says that the dimmer switch is sometimes built into the multi-function switch. Who is correct?

TASK D.5

A. A only

B. B only

C. Both A and B

D. Neither A nor B

Answer A is incorrect. Technician B is also correct.

Answer B is incorrect. Technician A is also correct.

Answer C is correct. Both Technicians are correct. The dash light dimmer is often built into the headlight switch. A rheostat is used to control the brightness of the dash lights by varying the voltage supplied to the dash lights. The dimmer switch is built into the multi-function switch on many late-model vehicles.

Answer D is incorrect. Both Technicians are correct.

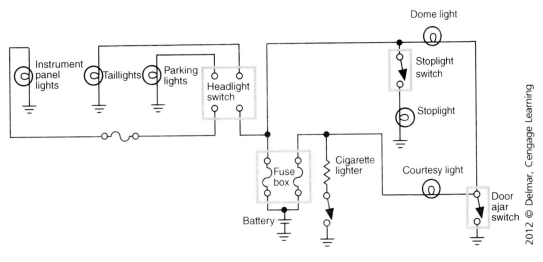

35. Referring to the figure above, the cigarette lighter fuse is blown in the circuit. The most likely result of this problem is:

A. The courtesy and dome lights come on dimly when you push in the lighter.

B. The stop and dome lights are completely inoperative.

TASK D.6, D.7

C. The parking lights, taillights, and instrument panel lights are inoperative.

D. The dome light will not work when the door is opened.

Answer A is correct. The courtesy light and dome light would work when the lighter is pushed in. The lighter would provide a path to ground.

Answer B is incorrect. The stop and dome lights receive power from a separate fuse and would not be affected by this problem.

Answer C is incorrect. The parking, instrument, and taillights receive power from a separate fuse and would not be affected by this problem.

Answer D is incorrect. The dome light sources power from the other fuse and would not be affected by this problem.

TASK A.7

36. A diode is being tested with a multimeter. Technician A says that a good diode test will read OL when the meter leads are connected in forward bias. Technician B says that a good diode will read approximately 0.5 volts when the meter leads are connected in reverse bias. Who is correct?

A. A only

B. B only

C. Both A and B

D. Neither A nor B

Answer A is incorrect. A good diode should read 0.5 to 0.7 volts when being forward biased with a multimeter.

Answer B is incorrect. A good diode should read OL when being reverse biased with a multimeter.

Answer C is incorrect. Neither Technician is correct.

Answer D is correct. Neither Technician is correct. Diodes can be tested with the diode tester on a digital multimeter. A good diode should read 0.5 to 0.7 volts when being forward biased with a multimeter. A good diode should read OL when being reverse biased with a multimeter. A shorted diode will produce a voltage drop when connected in forward or reverse bias. An open diode will measure OL when connected in forward or reverse bias.

TASK D.9

37. Technician A says that the hazard flasher on late-model vehicles can be replaced without disconnecting the vehicle battery. Technician B says that the turn signal flasher and hazard flasher are combined into one assembly on some late-model vehicles. Who is correct?

A. A only

B. B only

C. Both A and B

D. Neither A nor B

Answer A is incorrect. Technician B is also correct.

Answer B is incorrect. Technician A is also correct.

Answer C is correct. Both Technicians are correct. The vehicle battery would not have to be disconnected to replace the turn signal flasher assembly. Some vehicles have combined the turn signal flasher and the hazard flasher into one assembly.

Answer D is incorrect. Both Technicians are correct.

TASK D.11

38. All of the following tools can be used to test a trailer light cord EXCEPT:

A. Headlight with a jumper wire

B. Ammeter

C. 12 volt test light

D. Trailer cord test tool

Answer A is incorrect. A headlight used with a jumper wire could be used to test a trailer light cord operation. The headlight would draw enough current to easily find faults in the cord.

Answer B is correct. An ammeter is not typically used to test a trailer light cord. A tool that can sense voltage is needed to diagnose a trailer light cord.

Answer C is incorrect. A 12 volt test light could be used to test the trailer light cord. The lamp of the test light draws enough current to show the faults that may be present in the cord.

Answer D is incorrect. A trailer cord test tool is a good choice to use when testing the trailer light cord. This tool can be connected in series with the cord and the trailer and has an LED for each light circuit.

39. All of the following could cause an inaccurate gauge reading EXCEPT:

TASK E.3

    A. A defective ground at the sender unit
    B. High battery voltage
    C. A defective instrument voltage regulator (IVR)
    D. High resistance in the gauge wiring

Answer A is incorrect. A faulty ground at the sender unit could alter the circuit's overall resistance and throw off the gauge reading.

Answer B is correct. High battery voltage is compensated for by the IVR. Most magnetic gauges are not affected by varying voltage levels.

Answer C is incorrect. A faulty IVR could allow too much voltage to bi-metal gauges, resulting in an inaccurate reading.

Answer D is incorrect. Excessive resistance in the wiring will alter the circuit's overall resistance, throwing off the gauge readings.

40. What type of sending unit does the fuel gauge use?

TASK E.3

    A. Rheostat
    B. Transducer
    C. Photo resistor
    D. Thermistor

Answer A is correct. Fuel gauges typically use a rheostat or a potentiometer that connects to a float that rides on the fuel as it rises and falls.

Answer B is incorrect. A transducer is used on systems with varying pressure, such as the air conditioning (A/C) system.

Answer C is incorrect. A photo resistor is used as an input on automatic light systems, as well as some automatic temperature control systems.

Answer D is incorrect. A thermistor is used as a sending unit for the temperature gauges, as well as an input to measure temperature for several other electronic systems such as the engine, transmission, and the heating, ventilation, and air conditioning (HVAC) system.

41. Which of the following details would most likely be located on a wiring diagram?

TASK A.9

    A. Power and ground distribution for the circuit
    B. Location of the ground connection
    C. Updated factory information about pattern failures
    D. Flowchart for troubleshooting an electrical problem

Answer A is correct. A wiring diagram will provide details about how the power and ground are connected to the circuit.

Answer B is incorrect. A wiring diagram does not typically provide the location of electrical components such as a ground connection.

Answer C is incorrect. A wiring diagram does not typically provide updated factory information. Technical service bulletins provide updated factory information about pattern failures.

Answer D is incorrect. A wiring diagram does not provide any flowcharts for troubleshooting.

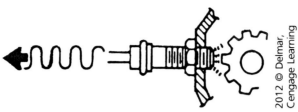

TASK E.5

42. Referring to the figure above, what can this type of sensor be used for?

    A. Speed sensing
    B. Pressure sensing
    C. Temperature sensing
    D. Level sensing

Answer A is correct. The magnetic pickup shown in the figure is used for shaft speed sensing, as used in tachometers and speedometers. The sensor contains a permanent magnet that creates a magnetic field that, when broken by a rotating gear, generates an AC voltage. The gauge assembly or computer module counts the impulses and then electronically computes the vehicle speed in miles per hour and the engine speed in revolutions per minute (RPM). This information is used to drive the speedometer and tachometer gauges to indicate the specific readings. Most of these speedometers and tachometers use the electromagnetic air core type of gauge.

Answer B is incorrect. The figure does not show a pressure sensor.

Answer C is incorrect. The figure does not show a temperature sensor.

Answer D is incorrect. The figure does not show a level sensor.

TASK E.8, E.9

43. Which of the following wiper problems would LEAST LIKELY cause slow wiper operation?

    A. Faulty wiper speed relay
    B. Blown wiper fuse
    C. Poor alternator output
    D. Excessively tight wiper linkage

Answer A is incorrect. A wiper relay with burned contacts could reduce the voltage being supplied to the wiper motor.

Answer B is correct. A blown wiper fuse would cause the windshield wipers to be totally inoperative.

Answer C is incorrect. Poor alternator output could cause slow wiper operation because of reduced amperage being supplied to the truck batteries.

Answer D is incorrect. Tight wiper linkage could cause slow wiper operation due to the increased physical resistance to motion.

44. The left-side heated mirror does not clear the mirror as well as the right-side heated mirror. Which of the following conditions would most likely cause this problem?

TASK E.11

    A. Faulty heated mirror switch
    B. Faulty right-side heated mirror
    C. Burned terminal at the driver's heated mirror electrical connector
    D. Weak alternator output

Answer A is incorrect. A faulty heated mirror switch would affect mirrors on both sides.

Answer B is incorrect. A faulty right-side heated mirror would not cause the left-side mirror to malfunction.

Answer C is correct. A burned electrical terminal at the left-side heated electrical connector would cause an unwanted voltage drop at that location; thus the left-side mirror would not clean as well.

Answer D is incorrect. Weak alternator output would have an effect on all electrical accessories.

45. Technician A says that the starter solenoid contacts can be tested by performing a voltage drop test across the "bat" and "motor" terminals when the starter is disengaged. Technician B says that the pull-in winding can be tested by measuring the resistance from the "S" terminal to the "motor" terminal. Who is correct?

TASK B.12

   A. A only

   B. B only

   C. Both A and B

   D. Neither A nor B

Answer A is incorrect. The starter contacts must be tested with the voltage drop test while the starter is engaged.

Answer B is correct. Only Technician B is correct. The pull-in winding can be tested with an ohmmeter connected from the "S" terminal to the "motor" terminal. This resistance is typically about 1 ohm.

Answer C is incorrect. Only Technician B is correct.

Answer D is incorrect. Technician B is correct.

46. Which of the following conditions would LEAST LIKELY cause all of the power window motors to be inoperative?

TASK E.15

   A. Faulty power window circuit breaker

   B. Bad ground at the master switch

   C. Binding power window regulator on the passenger side

   D. Blown power window fuse

Answer A is incorrect. A faulty power window circuit breaker would cause all power window motors to be inoperative.

Answer B is incorrect. A bad ground at the master switch would cause the power windows to be totally inoperative. The ground at the master switch is the only ground supplied to most power window circuits.

Answer C is correct. A binding passenger side window regulator would not likely cause all of the power window motors to be inoperative.

Answer D is incorrect. A blown power window fuse would cause all of the power window motors to be inoperative.

47. Which of the following conditions takes place when the truck is plugged into a shore power supply plug?

TASK E.16

   A. The truck is diagnosed for any engine electrical problems.

   B. The truck battery pack supplies power to the shore power system.

   C. The truck's electrical accessories receive power through the plug.

   D. The truck alternator supplies power to the shore power system.

Answer A is incorrect. A scan tool is needed to diagnose most engine electrical problems.

Answer B is incorrect. The truck battery pack does not supply power to the shore power system. The shore power system actually provides an electrical charge to the truck's battery pack.

Answer C is correct. The truck's electrical accessories receive power through the plug. In addition, the truck's battery pack receives an electrical charge while the truck is connected.

Answer D is incorrect. The truck alternator does not provide power to the shore power system plug.

TASK E.18

48. Which of the following steps would most likely be performed when replacing the power door lock actuator?

A. Remove the key lock tumbler.

B. Remove the door panel.

C. Remove the body control module (BCM).

D. Remove the power window regulator.

Answer A is incorrect. The key lock tumbler would not likely need to be removed during the power lock actuator replacement procedure.

Answer B is correct. The door panel would have to be removed to gain access to the power lock actuator.

Answer C is incorrect. The BCM would not have to be removed during the power lock actuator replacement.

Answer D is incorrect. The power window regulator would not have to be removed during the power lock actuator replacement.

TASK A.1

49. While diagnosing a truck with electronic engine management and a "no-start" complaint, Technician A says that a digital multimeter (DMM) could be used to check the electronic circuits. Technician B says that the DMM should be a high-impedance tool. Who is correct?

A. A only

B. B only

C. Both A and B

D. Neither A nor B

Answer A is incorrect. Technician B is also correct.

Answer B is incorrect. Technician A is also correct.

Answer C is correct. Both Technicians are correct. High-impedance DMMs are acceptable to be used on electronic circuits. Tools that do not have high impedance should not be used on electronic circuits.

Answer D is incorrect. Both Technicians are correct.

TASK E.20

50. Which of the following devices would LEAST LIKELY signal the ECM to actuate the electric fan on a late-model heavy-duty truck?

A. Trinary A/C switch

B. Turbo boost sensor

C. Coolant temperature sensor

D. A/C pressure sensor

Answer A is incorrect. A trinary switch is used in some heavy-duty trucks to provide three different functions. One of the functions of the trinary switch is to send a signal to turn on the engine fan when the A/C pressure reaches a certain point.

Answer B is correct. The turbo boost sensor is not typically used to control fan operation.

Answer C is incorrect. A coolant sensor is used on most heavy-duty trucks as an input to the control module based upon engine temperature.

Answer D is incorrect. An A/C pressure sensor is used on some heavy-duty trucks as an input to turn on the cooling fan when pressure rises to a certain level.

PREPARATION EXAM 5—ANSWER KEY

1.	D	18.	A	35.	D
2.	B	19.	A	36.	A
3.	C	20.	A	37.	B
4.	B	21.	A	38.	D
5.	B	22.	C	39.	C
6.	C	23.	C	40.	D
7.	C ,	24.	D	41.	C
8.	C	25.	C	42.	D
9.	B	26.	C	43.	D
10.	D	27.	D	44.	B
11.	C	28.	C	45.	D
12.	B	29.	B	46.	A
13.	B	30.	B	47.	B
14.	C	31.	C	48.	B
15.	A	32.	C	49.	C
16.	A	33.	A	50.	C
17.	C	34.	B		

PREPARATION EXAM 5—EXPLANATIONS

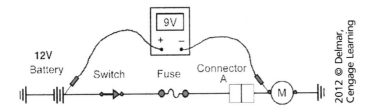

1. Referring to the figure above, the voltmeter in the circuit reads 9 volts. Which of the following conditions would be the most likely cause of this measurement?

    A. Wire repair with a larger than specified wire

    B. Faulty bulb

    C. Bad bulb ground

    D. Burned terminal connection at connector A

TASK A.1

    Answer A is incorrect. A larger wire used in a wire repair would not cause a voltage loss.

    Answer B is incorrect. There is no evidence that the bulb is faulty. The meter reads 9 volts being dropped across the switch, fuse, and connector.

    Answer C is incorrect. A bad bulb ground would not cause the 9 volt drop across the supply circuit.

    Answer D is correct. A burned terminal connection at Connector A could cause this problem. This problem could be isolated by performing a voltage drop test on Connector A.

TASK A.2

2. All of the following conditions can cause reduced current flow in an electrical circuit EXCEPT:

    A. Loose terminal connections

    B. Hot-side wire rubbing a metal component

    C. Corrosion inside the wire insulation

    D. A wire that is too small

    Answer A is incorrect. Loose terminal connections will reduce the current flow because of the added electrical resistance.

    Answer B is correct. A hot-side wire that rubs a metal component will cause a large increase in current flow and would likely open a circuit protection device.

    Answer C is incorrect. Wire corrosion will reduce the current flow because of the added electrical resistance.

    Answer D is incorrect. A wire that is too small will reduce the circuit's ability to carry the correct current because of the limitation of surface area to allow the electrons to flow.

TASK E.20

3. An electric fan on a late-model truck can be energized by which of the following methods?

    A. Viscous fluid switch

    B. Bi-metal spring switch

    C. Relay controlled by the engine control module (ECM)

    D. Vacuum thermal switch

    Answer A is incorrect. A mechanical fan sometimes uses a viscous-style device to control fan operation.

    Answer B is incorrect. A mechanical fan sometimes uses a bi-metal spring device to control fan operation.

    Answer C is correct. A relay that is controlled by the ECM is a common method to energize the electric cooling fan.

    Answer D is incorrect. Late-model trucks do not use thermal vacuum switches.

TASK A.4

4. Technician A says that a circuit with corrosion in the wiring will create a short circuit. Technician B says that a power wire that rubs a metal surface for a long period of time could create a short to ground. Who is correct?

    A. A only

    B. B only

    C. Both A and B

    D. Neither A nor B

    Answer A is incorrect. Corrosion in the wiring will not cause a short circuit.

    Answer B is correct. Only Technician B is correct. A power wire that rubs a metal surface could cause a short to ground. A circuit protection device would likely open if a short to ground occurs.

    Answer C is incorrect. Only Technician B is correct.

    Answer D is incorrect. Technician B is correct.

5. All of the following problems could cause excessive key-off battery drain EXCEPT:

TASK A.5

   A. Inverter switch stuck on

   B. Broken wire near the dome light socket

   C. Map light switch stuck closed

   D. Faulty rectifier bridge in the alternator

Answer A is incorrect. A stuck inverter switch could cause excessive key-off draw by causing the inverter to run continuously.

Answer B is correct. A broken wire near the dome light socket would cause an inoperative dome light. This problem would not cause excessive key-off battery drain because current does not flow in an open circuit.

Answer C is incorrect. A stuck closed map light switch could cause excessive key-off draw by causing the map light to run continuously.

Answer D is incorrect. A faulty rectifier bridge could cause excessive key-off draw by letting battery current flow into the alternator. The diodes will block any backfeeding when they are working correctly.

6. Which of the following devices would LEAST LIKELY be used as a circuit protection device?

TASK A.6

   A. Maxi-fuse

   B. Circuit breaker

   C. Relay

   D. Positive temperature coefficient (PTC) thermistor

Answer A is incorrect. Maxi-fuses have a metallic strip that burns up when the current flow in the circuit rises above the maxi-fuse rating. Once the fuse opens, it must be replaced to have continued electrical operation.

Answer B is incorrect. Circuit breakers are made of a bi-metallic strip that bends when it gets hot. High current flow in a circuit will cause the circuit breaker to heat up, which causes the device to open its contacts.

Answer C is correct. A relay is an electromagnetic switch that allows a large-current circuit to be controlled by a small-current circuit.

Answer D is incorrect. Some late-model trucks use PTC thermistors as a circuit protection device. PTC thermistors are electronic devices that have low resistance when they are at ambient temperature. The resistance of these devices increases as their temperature rises. This rising resistance limits current flow in the circuit they are protecting.

7. Technician A says that a loose drive belt could cause undercharging. Technician B says that undersized wiring between the alternator and the battery could cause undercharging. Who is correct?

TASK C.3, C.5

   A. A only

   B. B only

   C. Both A and B

   D. Neither A nor B

Answer A is incorrect. Technician B is also correct.

Answer B is incorrect. Technician A is also correct.

Answer C is correct. Both Technicians are correct. A loose drive belt or undersized wiring could result in an undercharging problem.

Answer D is incorrect. Both Technicians are correct.

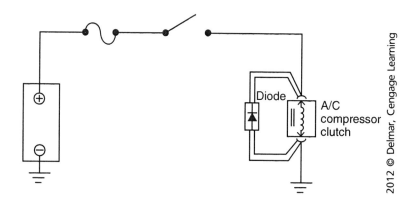

TASK A.7

8. Referring to the figure above, Technician A says that the air conditioner (A/C) compressor clutch coil is directly connected to ground. Technician B says that the diode in the circuit protects the components from the voltage spike that is created as the coil is de-energized. Who is correct?

A. A only

B. B only

C. Both A and B

D. Neither A nor B

Answer A is incorrect. Technician B is also correct.

Answer B is incorrect. Technician A is also correct.

Answer C is correct. Both Technicians are correct. The A/C compressor clutch coil is directly connected to ground. The diode in the schematic is used to provide a path for the voltage spike that is created when the coil is de-energized. Electromagnetic coils typically have either a diode or a resistor in parallel with them to handle these spikes.

Answer D is incorrect. Both Technicians are correct.

TASK B.8

9. Technician A says that a blown ignition fuse can cause low cranking voltage. Technician B says that a weak battery can cause low cranking voltage. Who is correct?

A. A only

B. B only

C. Both A and B

D. Neither A nor B

Answer A is incorrect. A blown ignition fuse would not likely cause low cranking voltage. If the fuse was part of the starter control circuit, the engine would not crank at all.

Answer B is correct. Only Technician B is correct. A weak battery in a set of batteries could cause low cranking voltage.

Answer C is incorrect. Only Technician B is correct.

Answer D is incorrect. Technician B is correct.

10. A wiring schematic is being used to diagnose a truck electrical problem. Technician A says that most schematics show the colors of the wires. Technician B says that most schematics are drawn with the power coming from the top of the picture. Who is correct?

TASK A.9

   A. A only

   B. B only

   C. Both A and B

   D. Neither A nor B

Answer A is incorrect. Technician B is also correct.

Answer B is incorrect. Technician A is also correct.

Answer C is correct. Both Technicians are correct. Most wiring schematic drawings show the wire colors labeled on the schematics. The colors are typically abbreviated with the letters in the color. For example, R=red, W=white, G=green. If the wire has a tracer color, the schematic will be labeled R/W=red with a white tracer or G/W=green with a white tracer. As a general rule, the drafters of wire schematics draw the picture with the power coming from top of the picture and ground coming from the bottom of the picture.

Answer D is incorrect. Both Technicians are correct.

11. When diagnosing a repeated flasher failure, Technician A says that a 100 millivolt voltage drop problem may be the cause. Technician B says that proper grounding of the trailer sockets may correct the problem. Who is correct?

TASK D.9

   A. A only

   B. B only

   C. Both A and B

   D. Neither A nor B

Answer A is incorrect. Even if the voltage drop were higher than the desired maximum of 100 millivolts, this would reduce current flow in the circuit, not increase it. This reduced current flow should not cause a flasher failure, although increased current flow could.

Answer B is incorrect. A poor ground at the trailer socket should cause reduced current flow, not increased current flow. For this reason, it should not cause a flasher unit to fail.

Answer C is incorrect. Neither Technician is correct.

Answer D is correct. Neither Technician is correct. The technician should look for conditions that would cause increased current flow, such as incorrect bulbs, short circuits, or improper mounting of the flasher.

12. Technician A states that the scan tool receives data from a connector located on the ECM. Technician B states that the scan tool connects to the data bus using the data link connector (DLC). Who is correct?

TASK A.10

   A. A only

   B. B only

   C. Both A and B

   D. Neither A nor B

Answer A is incorrect. There is no connector at the ECM to which the scan tool can be connected.

Answer B is correct. Only Technician B is correct. The scan tool connects to the data bus using a DLC which is located near the driver's-side area. Once connected, the scan tool is a bi-directional device that can read data, retrieve trouble codes, and send actuator commands to several systems.

Answer C is incorrect. Only Technician B is correct.

Answer D is incorrect. Technician B is correct.

TASK E.4

13. The indicator for the bright lights does not work, but the bright and dim headlights function correctly. Which of the following conditions would be the most likely cause of this problem?

A. Faulty headlight switch

B. Faulty bulb socket for the bright indicator

C. Faulty dimmer switch

D. Faulty multi-function switch

Answer A is incorrect. A headlight switch problem would cause all of the lights to malfunction.

Answer B is correct. A faulty bulb socket for the bright indicator could cause this problem.

Answer C is incorrect. A dimmer switch problem would cause a problem with the bright or dim headlights.

Answer D is incorrect. A multi-function switch problem would cause a problem with the bright or dim headlights.

TASK B.1, B.2

14. A truck is in the repair shop for a battery problem. Technician A says that the battery voltage should be 12.4 volts before performing a battery load test. Technician B says that a battery can be accurately tested with a digital tester if it has at least 12 volts. Who is correct?

A. A only

B. B only

C. Both A and B

D. Neither A nor B

Answer A is incorrect. Technician B is also correct.

Answer B is incorrect. Technician A is also correct.

Answer C is correct. Both Technicians are correct. A battery must have at least 12.4 volts before performing a valid load test. 12.4 volts represents a 75 percent charge level. A battery can be accurately tested with a digital battery tester if it has at least 12 volts. These testers perform a capacitance test on the battery and will let the technician know if the battery needs to be charged up or replaced.

Answer D is incorrect. Both Technicians are correct.

TASK B.3

15. A truck is being diagnosed for a battery problem. Technician A says when disconnecting battery cables, always disconnect the negative cable first. Technician B says when connecting battery cables, always connect the negative cable first. Who is correct?

A. A only

B. B only

C. Both A and B

D. Neither A nor B

Answer A is correct. Only Technician A is correct. Disconnecting the positive battery cable first will generate sparks and possibly a battery explosion if the wrench contacts ground.

Answer B is incorrect. Always reconnect battery cables in the reverse order of removal to avoid sparks and a possible explosion.

Answer C is incorrect. Only Technician A is correct.

Answer D is incorrect. Technician A is correct.

16. Technician A says that battery hold downs should always be installed to prevent batteries from bouncing, causing possible internal damage. Technician B states that a battery box need not be cleaned when replacing a battery because the case is insulated and the battery cannot discharge because of it. Who is correct?

TASK B.4

A. A only

B. B only

C. Both A and B

D. Neither A nor B

Answer A is correct. Only Technician A is correct. Batteries can be damaged internally if they are not properly secured in the battery tray. Improper mounting and hold down of the batteries can prematurely shorten their life. Although suspension systems have evolved to provide a smoother ride than ever before, most trucks are still prone to much flexing, vibration, and bouncing. These are further compounded by exposure to extreme environmental conditions. Any excessive bouncing and vibration can loosen cable connections, loosen the plate connections within the battery, and shake the active materials off the grid plates. Excessive bouncing can also allow electrolyte to escape, whether it is in liquid form or condensed vapors.

Answer B is incorrect. Batteries can self-discharge through any accumulated corrosion and moisture buildup across the top. A dirty battery tray will accelerate this process.

Answer C is incorrect. Only Technician A is correct.

Answer D is incorrect. Technician A is correct.

17. A truck with a charging problem is being repaired. The technician finds that the charging output wire received damage and needs to be repaired. Technician A says to use weather-resistant connectors when making wire repairs in the engine compartment. Technician B says that solder and heat shrink is an acceptable method of wire repair in the engine compartment. Who is correct?

TASK C.7

A. A only

B. B only

C. Both A and B

D. Neither A nor B

Answer A is incorrect. Technician B is also correct.

Answer B is incorrect. Technician A is also correct.

Answer C is correct. Both Technicians are correct. Water-resistant wire repair methods should always be used when making repairs in the engine compartment. Weather-resistant connectors, as well as solder and heat shrink, are both acceptable ways to perform wire repairs in the engine compartment.

Answer D is incorrect. Both Technicians are correct.

TASK B.9

18. A starter circuit voltage drop test checks everything EXCEPT:

    A. Battery voltage
    B. Resistance in the positive battery cable
    C. Resistance in the negative battery cable
    D. Condition of the solenoid internal contacts

    Answer A is correct. A starter voltage drop test does not check the condition of a battery nor its state of charge, only the resistance to current flow in the various connections and cables.

    Answer B is incorrect. Resistance in the positive battery cables can be checked during a starter circuit voltage drop test.

    Answer C is incorrect. Resistance in the negative battery cables can be checked during a starter circuit voltage drop test.

    Answer D is incorrect. Resistance in the solenoid internal contacts can be checked during a starter circuit voltage drop test.

TASK B.12

19. Which of the following functions would be LEAST LIKELY performed by the starter solenoid?

    A. Preventing the armature from overspinning
    B. Pushing the drive gear out to the flywheel
    C. Connecting the "bat" terminal to the "motor" terminal
    D. Providing a path for high current to flow

    Answer A is correct. The starter solenoid has no control over the speed of the armature.

    Answer B is incorrect. The starter solenoid creates linear movement to push the drive gear into the flywheel.

    Answer C is incorrect. The starter solenoid acts as a switch to connect the "bat" terminal to the "motor" terminal when the starter is engaged.

    Answer D is incorrect. The starter solenoid provides an electrical path for high current flow through the contacts and into the starter housing.

TASK B.14

20. A starter is being removed from a truck. Technician A says that the negative battery cable should be removed prior to disconnecting the electrical connections at the starter. Technician B says that the fasteners should be removed prior to removing the electrical connections at the starter. Who is correct?

    A. A only
    B. B only
    C. Both A and B
    D. Neither A nor B

    Answer A is correct. Only Technician A is correct. It is advisable to always disconnect the battery prior to removing the starter motor. It is safer to remove the negative cable first due to the possibility of letting the wrench touch metal while connected to the cable end.

    Answer B is incorrect. The electrical connections should be removed before removing the fasteners, if possible, so that the starter does not hang by the connecting wires.

    Answer C is incorrect. Only Technician A is correct.

    Answer D is incorrect. Technician A is correct.

21. What is the function of the park circuit in the windshield wiper system?

TASK E.9

    A. Return the wiper blades to the start position.

    B. Shut down the wiper motor in case of overheating.

    C. Help keep the wiper blades synchronized.

    D. Stop the wiper motor in case of a low-voltage problem.

Answer A is correct. The function of the park switch is to return the wipers to the start position. One such system incorporates a set of park contacts into the motor assembly that operates off a cam or latch arm on the motor gear. The park switch changes position with each revolution of the motor. The switch remains in the run position for approximately 90 percent of the revolution and in the park position for the remaining 10 percent of the revolution. The operation of the wiper motor is not affected until the wiper control switch is placed in the park position.

Answer B is incorrect. The thermal overload switch shuts down the wiper motor in case of overheating.

Answer C is incorrect. It is the function of the linkage to synchronize the blades (single-motor systems).

Answer D is incorrect. There is no low-voltage protection for a wiper motor.

22. A truck is being diagnosed for a slow crank. Smoke is noticed at the starter solenoid when the truck is in the crank mode. Technician A says that a loose attachment nut at the starter solenoid could be the cause. Technician B says that the connections at the starter solenoid should be checked for burned contacts. Who is correct?

TASK B.13

    A. A only

    B. B only

    C. Both A and B

    D. Neither A nor B

Answer A is incorrect. Technician B is also correct.

Answer B is incorrect. Technician A is also correct.

Answer C is correct. Both Technicians are correct. A loose connection at the starter solenoid could cause smoke in that area when the truck is being cranked. Burned electrical contacts would result from the loose attachment nut at the solenoid.

Answer D is incorrect. Both Technicians are correct.

23. A truck is being diagnosed for a starting problem. Technician A states that weak batteries can cause high current draw. Technician B states that faulty starter bushings can cause high current draw. Who is correct?

TASK B.15

    A. A only

    B. B only

    C. Both A and B

    D. Neither A nor B

Answer A is incorrect. Technician B is also correct.

Answer B is incorrect. Technician A is also correct.

Answer C is correct. Both Technicians are correct. Weak batteries and faulty starter bushings can cause high current draw accompanied with slow cranking speed.

Answer D is incorrect. Both Technicians are correct.

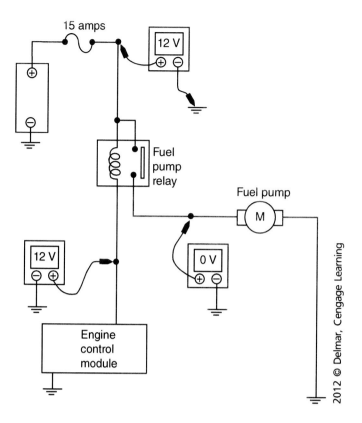

15 amps

12 V

Fuel pump relay

Fuel pump

M

12 V

0 V

Engine control module

2012 © Delmar, Cengage Learning

TASK A.8

24. Referring to the fuel pump circuit in the figure above, the fuel pump is inoperative and the voltage readings were taken immediately after the circuit was closed. Technician A says that the fuel pump is likely defective. Technician B says that the relay coil could be open. Who is correct?

A. A only

B. B only

C. Both A and B

D. Neither A nor B

Answer A is incorrect. The fuel pump is not being supplied any voltage in the circuit.

Answer B is incorrect. The voltage near the control module is measured at 12 volts, which disproves the possibility of the relay coil being open. If the coil was open, there would not be any voltage present near the control module.

Answer C is incorrect. Neither Technician is correct.

Answer D is correct. Neither Technician is correct. The most likely problem in the circuit is a problem in the ECM due to the 12 volts being present on the ground side of the relay coil. If the ECM were providing ground to the relay coil, then the voltage on the ground side of the coil would be near zero.

25. A truck's charging system produces 55 amps during a charging performance test. The charging specification for the truck is 110 amps. Technician A says that a full-field test should be performed to see if the charging performance improves. Technician B says that the charging circuit voltage drop should be tested. Who is correct?

TASK C.2

  A. A only
  B. B only
  C. Both A and B
  D. Neither A nor B

Answer A is incorrect. Technician B is also correct.

Answer B is incorrect. Technician A is also correct.

Answer C is correct. Both Technicians are correct. A full-field test should be run when an alternator fails to produce within 10 percent of its rated output during a performance test. This test by-passes the voltage regulator. If the charge rate comes up to specifications during this test, then the voltage regulator needs to be replaced. A voltage drop test is also a valid test to run when an alternator does not charge within 10 percent of its rated output.

Answer D is incorrect. Both Technicians are correct.

26. Which of the following devices would LEAST LIKELY be used as a spike suppression device for an electromagnetic coil?

TASK A.7

  A. Resistor
  B. Diode
  C. Transistor
  D. Zener diode

Answer A is incorrect. A resistor is sometimes used as a spike suppression device for an electromagnetic coil.

Answer B is incorrect. A diode is sometimes used as a spike suppression device for an electromagnetic coil.

Answer C is correct. A transistor is not typically used as a spike suppression device for an electromagnetic coil. A transistor is a solid-state electronic component that is used as a switch or as a current multiplier in control modules.

Answer D is incorrect. A zener diode is sometimes used as a spike suppression device for an electromagnetic coil.

27. An alternator output test is being performed. Technician A uses only a voltmeter connected across the battery positive terminal and negative terminal while the engine is running. Technician B says a carbon pile is not needed since the engine is already running. Who is correct?

TASK C.4

  A. A only
  B. B only
  C. Both A and B
  D. Neither A nor B

Answer A is incorrect. An ammeter is also needed to determine the current output of the alternator.

Answer B is incorrect. A carbon pile is needed to load the system to force maximum alternator output.

Answer C is incorrect. Neither Technician is correct.

Answer D is correct. Neither Technician is correct. Just using a voltmeter to test the alternator is not very effective because current output is the most accurate way of testing an alternator. A tester with a carbon pile is the best way to fully load the electrical system and cause the alternator to charge at full capacity.

TASK C.4

28. A technician is testing alternator output. After starting the engine the test reveals that current output from the alternator slowly decreases the longer the engine runs. What can this mean?

   A. Alternator output is marginal; discontinue the test.
   B. The alternator drive belt is probably slipping on the pulley.
   C. The battery is slowly recovering to capacity.
   D. The diodes in the alternator are heating up and starting to fail.

   Answer A is incorrect. This action is normal. The battery is being replenished quickly, so the charging current going into it will steadily decrease as time goes on.

   Answer B is incorrect. A slipping drive belt should produce a consistent current output from the alternator.

   Answer C is correct. This is a normal occurrence as a battery is recharged.

   Answer D is incorrect. Diodes typically do not fail due to heat stress.

TASK C.5

29. A truck is being diagnosed for a charging problem. The alternator only charges at 12.4 volts. A voltage drop test is performed on the charging output wire and 1.8 volts are measured. Technician A says that a blown fusible link in the output circuit could be the cause. Technician B says that a loose nut at the charging output connector could be the cause. Who is correct?

   A. A only
   B. B only
   C. Both A and B
   D. Neither A nor B

   Answer A is incorrect. A blown fusible link would have produced a higher voltage drop.

   Answer B is correct. Only Technician B is correct. A loose nut at the charging output connector could cause the excessive voltage drop in the charging circuit.

   Answer C is incorrect. Only Technician B is correct.

   Answer D is incorrect. Technician B is correct.

TASK C.6

30. Which of the following practices would be LEAST LIKELY followed when replacing the alternator on a heavy-duty truck?

   A. Use a box-end wrench to loosen the charging output nut.
   B. Remove the brushes from the old alternator and install them into the replacement alternator.
   C. Remove the drive belt prior to removing the alternator.
   D. Disconnect the negative battery cable prior to removing the alternator.

   Answer A is incorrect. It is a good practice to use a box-end wrench when loosening the charging output nut to prevent damage to the nut from rounding the corners.

   Answer B is correct. It is not necessary to remove the brushes from the old alternator during the replacement process. The replacement alternator should have new brushes that work well.

   Answer C is incorrect. It is a good practice to remove the drive belt prior to removing the alternator. The belt should be closely inspected for wear and damage during the replacement process.

   Answer D is incorrect. It is a good practice to remove the negative battery cable prior to removing the alternator to prevent accidentally grounding out the charging output circuit during the repair.

31. Technician A says that the low voltage disconnect (LVD) module is connected in series between the truck battery pack and the truck power distribution center. Technician B says that the LVD system will turn off battery power when the battery voltage level drops below 10.4 volts. Who is correct?

TASK B.7

    A. A only

    B. B only

    C. Both A and B

    D. Neither A nor B

Answer A is incorrect. Technician B is also correct.

Answer B is incorrect. Technician A is also correct.

Answer C is correct. Both Technicians are correct. The LVD module is installed between the battery pack and the power distribution center. This system protects the truck electrical system from totally discharging the starting batteries. This system disconnects the batteries when the voltage level drops to approximately 10.4 volts.

Answer D is incorrect. Both Technicians are correct.

32. Technician A says that an overcharging alternator can cause lights that are brighter than normal. Technician B states that poor chassis grounds usually cause dim lights. Who is correct?

TASK D.1

    A. A only

    B. B only

    C. Both A and B

    D. Neither A nor B

Answer A is incorrect. Technician B is also correct.

Answer B is incorrect. Technician A is also correct.

Answer C is correct. Both Technicians are correct. An alternator that is overcharging will cause excessive system voltage that will increase the current flow through the lights, causing them to be brighter than normal. Poor chassis grounds are usually the cause of dim lights due to rust and corrosion at mounting points.

Answer D is incorrect. Both Technicians are correct.

33. Technician A says that it is a good idea to coat the prongs and base of a new sealed beam light assembly with dielectric grease before installing to prevent corrosion. Technician B says that white lithium grease can be used instead of dielectric grease. Who is correct?

TASK D.2

    A. A only

    B. B only

    C. Both A and B

    D. Neither A nor B

Answer A is correct. Only Technician A is correct. Dielectric lubricant is the recommended lubricant for almost all electrical connections.

Answer B is incorrect. Lithium grease is not recommended for electrical connections.

Answer C is incorrect. Only Technician A is correct.

Answer D is incorrect. Technician A is correct.

TASK A.5,
D.6

34. The courtesy lights stay on continuously and have caused the batteries to be discharged while the truck is parked for long periods of time. Technician A says that the courtesy light switch could have a bad connection. Technician B says that a door ajar switch could be shorted. Who is correct?

   A.   A only

   B.   B only

   C.   Both A and B

   D.   Neither A nor B

Answer A is incorrect. A bad connection at the courtesy light switch would likely cause the lights to be dim or possibly be totally inoperative.

Answer B is correct. Only Technician B is correct. A shorted door ajar switch could cause the courtesy lights to stay on continuously.

Answer C is incorrect. Only Technician B is correct.

Answer D is incorrect. Technician B is correct.

TASK D.8

35. A truck is being diagnosed for inoperative turn signals. All of the hazard lights work correctly. Technician A says that a blown turn signal bulb could be the cause. Technician B says that an open ground connection at the right-rear lamp socket could be the cause. Who is correct?

   A.   A only

   B.   B only

   C.   Both A and B

   D.   Neither A nor B

Answer A is incorrect. The turn signals and the hazard lights use the same bulbs, so a blown bulb would not be the cause of this problem.

Answer B is incorrect. The turn signals and the hazard lights use the same bulbs, so an open ground at the right-rear lamp socket would not be the cause of this problem.

Answer C is incorrect. Neither Technician is correct.

Answer D is correct. Neither Technician is correct. This problem has to be something that is common only to the turn signal system. Both of the examples would have caused the hazard lights to not burn correctly. The most likely causes of this problem would be a faulty turn signal flasher or a faulty turn signal switch.

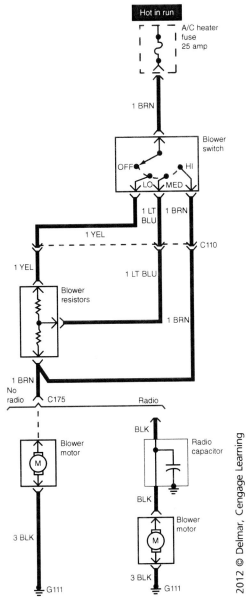

36. Referring to the figure above, which of the following statements best describes the wiring schematic?

A. Power is sent to the light blue wire when the blower switch is set to medium speed.

B. Power is sent to the yellow wire when the blower switch is set to the high speed.

C. Power is sent to the brown wire when the blower switch is set to the low speed.

D. The A/C heater fuse supplies ground to the blower motor circuit.

TASK A.9

Answer A is correct. Power is sent to the light blue wire when the blower motor is set to the medium speed. This action forces the current to pass through one of the resistors and then through the blower motor, which causes the blower motor to run at medium speed.

Answer B is incorrect. Power is sent to the yellow wire when the blower switch is set to the low speed. This forces the current through both of the resistors and then to the blower motor, which causes the blower motor to run slowly.

Answer C is incorrect. Power is sent to the brown wire when the switch is set to the high speed. This action by-passes the resistors and causes the blower motor to run on high speed.

Answer D is incorrect. The A/C heater fuse supplies power to the blower motor circuit.

TASK D.10

37. The back-up lights are inoperative on a truck. Technician A says that a faulty brake switch could be the cause. Technician B says that a stuck open back-up lamp switch could be the cause. Who is correct?

A. A only

B. B only

C. Both A and B

D. Neither A nor B

Answer A is incorrect. The back-up light circuit does not pass through the brake switch.

Answer B is correct. Only Technician B is correct. A stuck open back-up lamp switch could cause the back-up lights to be inoperative. A good switch should open and close as it is operated.

Answer C is incorrect. Only Technician B is correct.

Answer D is incorrect. Technician B is correct.

TASK A.12

38. Technician A says that the main data link connector (DLC) is a square 9-pin connector. Technician B says that the scan tool should be used to engage flash code diagnostics on the truck. Who is correct?

A. A only

B. B only

C. Both A and B

D. Neither A nor B

Answer A is incorrect. The data connector on late-model trucks is a round 9-pin design.

Answer B is incorrect. A scan is not needed to enter flash code diagnostics on most vehicles.

Answer C is incorrect. Neither Technician is correct.

Answer D is correct. Neither Technician is correct. Late-model trucks use a round 9-pin data connector that is located somewhere inside the cab. Flash code diagnostics can be engaged by signaling the computer. Some systems have a switch that needs to be depressed to enter this mode.

TASK E.2

39. All of the following types of electronic devices are used as inputs to electronic gauge assemblies EXCEPT:

A. Thermistor

B. Piezo resistor

C. Diode

D. Rheostat

Answer A is incorrect. Thermistors are used as inputs to electronic temperature gauges.

Answer B is incorrect. Piezo resistors are used as inputs to electronic oil pressure gauges.

Answer C is correct. Diodes are not used as inputs to electronic gauges assemblies. Diodes are used widely inside computers, as well as in clamping devices for electromagnetic coils.

Answer D is incorrect. Rheostats are used as inputs to many electronic fuel gauges.

40. On a truck with an electronic dash display, all of the gauge needles sweep from left to right immediately after turning the key on. Technician A says that this may indicate a fault with the instrument panel. Technician B states that this is due to high battery voltage. Who is correct?

TASK E.2

  A. A only
  B. B only
  C. Both A and B
  D. Neither A nor B

Answer A is incorrect. This gauge action occurs during an instrument self-test that verifies operation of all the gauges.

Answer B is incorrect. It is not very likely that battery voltage would be too high when the key is first turned on. Even if the battery voltage were too high, the stated gauge action is normal.

Answer C is incorrect. Neither Technician is correct.

Answer D is correct. Neither Technician is correct. The action described is normal for an electronic instrument panel. Most of today's computerized vehicles have the capability of performing an initial power-up or self-test of the speedometer/tachometer unit. The absence of a signal is usually indicated by the gauge pointer moving to its minimum reading.

41. All of the following methods can be used to retrieve codes from a truck's on-board computer EXCEPT:

TASK A.11, A.12

  A. Laptop-based scan tool
  B. Self-contained scan tool
  C. Technical service bulletin
  D. Flash code diagnostics

Answer A is incorrect. A laptop-based scan tool is commonly used to retrieve trouble codes from a computer system on a truck.

Answer B is incorrect. A self-contained scan tool is sometimes used to retrieve trouble codes from a computer system on a truck.

Answer C is correct. A technical service bulletin would not be used to retrieve trouble codes from a computer system on a truck. Technical service bulletins are documents that are created by the manufacturer to assist the technicians in repairing pattern failures on trucks.

Answer D is incorrect. Flash-code diagnostics are often used to retrieve trouble codes from the computer systems on the truck. This process is started by the technician depressing a switch or otherwise signaling the computer to display the fault codes with a flashing indicator.

42. The horn blows intermittently on a heavy-duty truck. Which of the following conditions would be the LEAST LIKELY cause of this problem?

TASK E.6

  A. Wire rubbing a ground near the base of the steering column
  B. Sticking horn relay
  C. Faulty horn switch
  D. Horn fault

Answer A is incorrect. A wire rubbing a ground near the steering column could cause the horn to blow intermittently.

Answer B is incorrect. A sticking horn relay could cause the horn to blow intermittently.

Answer C is incorrect. A faulty horn switch could cause the horn to blow intermittently.

Answer D is correct. A horn fault would not likely cause an intermittent condition of the horn sounding.

TASK E.7

43. An electric horn on a medium-duty truck operates intermittently. Which of the following could be the cause?

    A. Blown fuse
    B. No power to relay
    C. Open in horn button circuit
    D. Defective horn relay

Answer A is incorrect. A blown fuse would make the horn totally inoperative, not intermittently inoperative.

Answer B is incorrect. No power to the relay would make the horn totally inoperative, not intermittently inoperative.

Answer C is incorrect. An open in the horn button circuit would make the horn totally inoperative, not intermittently inoperative.

Answer D is correct. A defective horn relay could cause intermittent horn operation if its action is erratic. Several factors that can affect the sound quality of a horn or a multiple horn system are: damaged diaphragms, poor ground connections, improperly adjusted horns, or excessive circuit resistance.

TASK E.8

44. Which one of the following problems would be the most likely cause for inoperative windshield wipers?

    A. Stuck closed wiper switch
    B. Tripped thermal overload protector
    C. Closed circuit breaker
    D. High resistance in the motor wiring

Answer A is incorrect. A stuck closed wiper switch would cause the wipers to run constantly.

Answer B is correct. A tripped thermal overload protector would open the circuit and prevent current flow to the motor.

Answer C is incorrect. A closed circuit breaker would not cause a problem for the windshield wipers. A closed circuit breaker would allow the necessary current to flow to the wipers.

Answer D is incorrect. High resistance in the wiring should only slow the motor down, not stop it.

TASK B.14

45. Technician A says that the courtesy fuse should be removed prior to disconnecting the electrical connections at the starter. Technician B says that the starter can be supported by the connecting electrical wires without any expected damage to the wires. Who is correct?

    A. A only
    B. B only
    C. Both A and B
    D. Neither A nor B

Answer A is incorrect. It is not necessary to remove the courtesy fuse when unhooking the starter.

Answer B is incorrect. It is not acceptable to let the starter hang by the electrical wires.

Answer C is incorrect. Neither Technician is correct.

Answer D is correct. Neither Technician is correct. The negative battery cable should be disconnected prior to disconnecting the electrical wires at the starter. The starter should never be supported by just the electrical wires.

46. Which of the following conditions would be most likely to cause the power windows to be inoperative?

TASK E.15

    A. Missing ground connection to the master switch

    B. Faulty power lock switch

    C. Binding power window regulator

    D. Loose door panel

Answer A is correct. A missing ground at the master switch would cause all of the power window motors to be inoperative, because this is the only ground supplied to the power window circuit.

Answer B is incorrect. A faulty power lock switch would have no affect on the power window operation.

Answer C is incorrect. A binding power window regulator would only cause problems at one location.

Answer D is incorrect. A loose door panel would not likely cause a problem for the power window system.

47. Which of the following conditions takes place when a truck is plugged into a shore power supply plug?

TASK E.16

    A. The truck is diagnosed for any engine electrical problems.

    B. The battery pack receives an electrical charge.

    C. The truck battery pack supplies power to the shore power system plug.

    D. The truck alternator supplies power to the shore power system plug.

Answer A is incorrect. A scan tool is needed to diagnose most engine electrical problems.

Answer B is correct. The truck battery pack receives an electrical charge while the truck is connected to a shore power supply plug. In addition, the truck's accessories receive power from the shore power plug.

Answer C is incorrect. The truck battery does not supply power to the shore power system plug. The shore power can charge the truck batteries when it is connected to the supply plug at a truck stop.

Answer D is incorrect. The truck alternator does not supply power to the shore power system plug.

48. What kind of component is most likely used as a power lock actuator?

TASK E.18

    A. Vacuum actuator

    B. Solenoid

    C. Electromagnetic motor

    D. Relay

Answer A is incorrect. Although vacuum is used in several systems on light-duty gasoline trucks, diesel trucks rarely use vacuum to actuate any systems; vacuum is not naturally produced in a diesel engine.

Answer B is correct. Many power door lock actuators are solenoids. A solenoid can be used in many applications where linear movement is needed. A solenoid is also bi-directional when polarity is changed.

Answer C is incorrect. Electromagnetic motors make good starter motors but are not bi-directional, so they would not work for power lock actuators. Permanent magnet (PM) motors are sometimes used since they are bi-directional.

Answer D is incorrect. A relay is very effective when used as a switching device, but it cannot be used to provide any linear movement.

TASK A.3

49. Technician A says that power must be turned off in the circuit before using an ohmmeter to make a measurement. Technician B says that the ohmmeter applies a small amount of voltage to the circuit to calculate resistance. Who is correct?

    A. A only
    B. B only
    C. Both A and B
    D. Neither A nor B

    Answer A is incorrect. Technician B is also correct.

    Answer B is incorrect. Technician A is also correct.

    Answer C is correct. Both Technicians are correct. An ohmmeter should never be connected to a live electrical circuit because it produces its own voltage when configured as an ohmmeter. Connecting an ohmmeter to a live circuit will likely damage the meter.

    Answer D is incorrect. Both Technicians are correct.

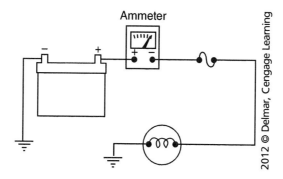

2012 © Delmar, Cengage Learning

TASK A.2

50. Referring to the figure above, the ammeter indicates that current flow through the bulb is higher than specified. Which of the following could be the cause of this high current?

    A. The fuse has an open circuit.
    B. The battery voltage is low.
    C. The light bulb filament is shorted.
    D. The bulb filament has high resistance.

    Answer A is incorrect. If the fuse were open, there would be no current flow at all through the circuit.

    Answer B is incorrect. Reduced battery voltage will cause a reduction in current flow if the resistance of the load remains the same (Ohm's Law).

    Answer C is correct. A shorted bulb filament will cause reduced resistance in the bulb that will cause current flow to increase if the battery voltage remains constant (Ohm's Law). An ammeter is used to measure current flow. The meter is placed in series so it becomes part of the circuit, unless an inductive ammeter probe is used. Either is a good indicator of current draw.

    Answer D is incorrect. An increase in bulb resistance will cause a reduction in current flow if the battery voltage remains constant (Ohm's Law).

PREPARATION EXAM 6—ANSWER KEY

1.	C	18.	D	35.	B
2.	D	19.	A	36.	C
3.	A	20.	A	37.	D
4.	C	21.	B	38.	B
5.	B	22.	C	39.	B
6.	C	23.	B	40.	C
7.	D	24.	B	41.	C
8.	C	25.	B	42.	C
9.	C	26.	D	43.	A
10.	C	27.	D	44.	D
11.	B	28.	B	45.	D
12.	C	29.	B	46.	B
13.	D	30.	D	47.	C
14.	D	31.	D	48.	D
15.	B	32.	C	49.	D
16.	B	33.	A	50.	D
17.	C	34.	C		

PREPARATION EXAM 6—EXPLANATIONS

1. A scan tool is connected to the engine computer of a late-model truck with a check engine light that stays on. Technician A says that the system should be checked for diagnostic trouble codes (DTCs). Technician B says that the data list should be checked for irregular readings. Who is correct?

TASK A.11, A.12

A. A only

B. B only

C. Both A and B

D. Neither A nor B

Answer A is incorrect. Technician B is also correct.

Answer B is incorrect. Technician A is also correct.

Answer C is correct. Both Technicians are correct. The trouble codes should be checked when troubleshooting a check engine light. It is also advisable to view the data list to check for irregular readings from the sensors and switches.

Answer D is incorrect. Both Technicians are correct.

TASK A.2

2. How is an ammeter connected in order to read live current?

   A. In parallel with the circuit with the power turned off

   B. In parallel with the circuit with the power turned on

   C. In series with the circuit with the power turned off

   D. In series with the circuit with the power turned on

Answer A is incorrect. Connecting the ammeter in parallel with the circuit and with the power turned off is the correct way to make a resistance measurement with an ohmmeter.

Answer B is incorrect. Connecting the ammeter in parallel with the circuit and with the power turned on is the correct way to make a voltage measurement with a voltmeter.

Answer C is incorrect. The power would have to be turned on in order to read a live current.

Answer D is correct. The ammeter has to be connected in series with an electrical circuit with the power turned on to read a live current. Care should be taken after a current measurement is made to move the red digital multimeter (DMM) lead to the volt slot before making other measurements. If the lead is left in the amp slot on the meter, it is easy to blow the internal DMM fuse if the leads accidentally touch a live circuit in parallel.

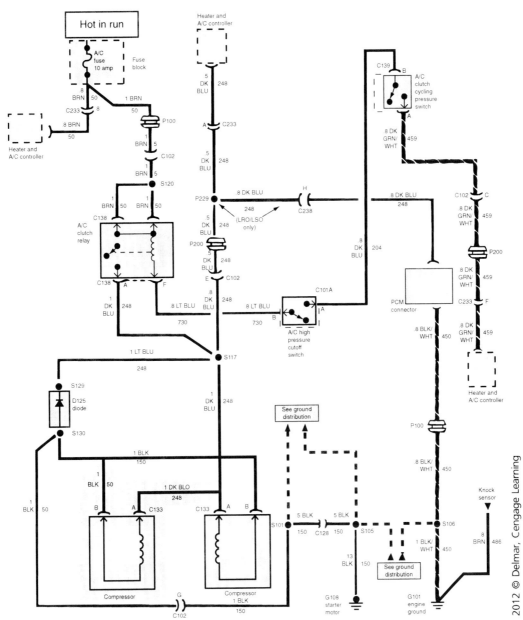

3. Referring to the figure above, this electrical schematic is being examined by two technicians. Technician A says that the schematic shows the heater and air conditioner (A/C) controller circuit. Technician B says that this circuit operates with circuit #50 (1 brn) being open. Who is correct?

TASK A.9,
E.12

A. A only

B. B only

C. Both A and B

D. Neither A nor B

Answer A is correct. Only Technician A is correct. The schematic shows both heater and A/C controller components.

Answer B is incorrect. An open in circuit #50 (1 brn) would prevent the A/C clutch relay from operating.

Answer C is incorrect. Only Technician A is correct.

Answer D is incorrect. Technician A is correct.

.TASK A.5

4. Technician A says that a stuck closed dome light switch could cause excessive key-off drain. Technician B says that a faulty starter solenoid could cause excessive key-off drain. Who is correct?

A. A only

B. B only

C. Both A and B

D. Neither A nor B

Answer A is incorrect. Technician B is also correct.

Answer B is incorrect. Technician A is also correct.

Answer C is correct. Both Technicians are correct. A stuck closed dome light switch would cause the dome lights to remain illuminated at all times which would create excessive key-off drain. A starter solenoid that is internally shorted will cause excessive key-off drain because the positive battery cable connects to the starter solenoid.

Answer D is incorrect. Both Technicians are correct.

TASK A.5, B.3

5. A truck with a key-off battery drain problem is being diagnosed. Technician A states that having zero key-off battery drain is common on a truck equipped with multiple control modules. Technician B says battery drain can be caused by excess moisture on top of the battery. Who is correct?

A. A only

B. B only

C. Both A and B

D. Neither A nor B

Answer A is incorrect. Trucks that are equipped with multiple control modules will usually have some measurable key-off draw. However, these modules should only draw a few milliamps when the ignition switch is in the off position.

Answer B is correct. Only Technician B is correct. Batteries can sometimes self-discharge when there is excess moisture on top of the battery. Care should be taken to keep the battery cases clean and dry if possible.

Answer C is incorrect. Only Technician B is correct.

Answer D is incorrect. Technician B is correct.

TASK A.6

6. A fusible link needs to be replaced. Technician A says that the batteries should be disconnected first. Technician B says that the same gauge fuse link wire should be used in the repair. Who is correct?

A. A only

B. B only

C. Both A and B

D. Neither A nor B

Answer A is incorrect. Technician B is also correct.

Answer B is incorrect. Technician A is also correct.

Answer C is correct. Both Technicians are correct. It is advisable to disconnect the batteries prior to replacing a fusible link to prevent possible damage to the truck from shorting the wire to ground while making the repair. After verifying that the correct fusible link was installed, the technician should always use the same size that was removed.

Answer D is incorrect. Both Technicians are correct.

7. An alternator is overcharging. Technician A says that this can only be caused by a defective voltage regulator. Technician B states that excessive resistance in the charging circuit wiring can cause this overcharging. Who is correct?

TASK C.2

    A. A only
    B. B only
    C. Both A and B
    D. Neither A nor B

Answer A is incorrect. While a defective voltage regulator is the most likely cause, alternator overcharging can also be caused by a fault in the sense circuit on systems with external regulators.

Answer B is incorrect. Excessive resistance in the charging circuit wiring should cause undercharging, not overcharging.

Answer C is incorrect. Neither Technician is correct.

Answer D is correct. Neither Technician is correct. A defective voltage regulator is a very likely cause of an overcharging alternator. However, another possible cause could be a resistance problem in the regulator sense circuit. A resistance problem in the charging circuit would result in undercharging, not overcharging.

8. Technician A says that resistors are sometimes used as spike suppression devices in relays. Technician B says that diodes are sometimes used as spike suppression devices in relays. Who is correct?

TASK A.7

    A. A only
    B. B only
    C. Both A and B
    D. Neither A nor B

Answer A is incorrect. Technician B is also correct.

Answer B is incorrect. Technician A is also correct.

Answer C is correct. Both Technicians are correct. Electromagnetic coils typically have either a diode or a resistor in parallel with them to handle the voltage spikes that are created when the coil is de-energized.

Answer D is incorrect. Both Technicians are correct.

9. Which of the following conditions would LEAST LIKELY cause a slow crank problem?

TASK B.15

    A. Mistimed engine
    B. Worn starter brushes
    C. Battery with a higher CCA rating than original specifications
    D. Engine with tight main bearings

Answer A is incorrect. A mistimed engine could cause a slow crank problem due to the ignition system firing at the wrong time or if the crank/cam timing were incorrect.

Answer B is incorrect. Worn starter brushes could cause a slow crank problem due to the increased electrical resistance.

Answer C is correct. A battery with a higher CCA rating would not cause a slow crank problem. Batteries with higher CCA ratings can be used if the physical dimensions allow it.

Answer D is incorrect. An engine with tight main bearings could cause a slow crank problem due to the increased physical resistance.

TASK A.8

10. Technician A says that a solenoid can stick in the applied position if the return spring breaks. Technician B says that a solenoid has a coil that can be tested with an ohmmeter for the correct resistance. Who is correct?

   A. A only
   B. B only
   C. Both A and B
   D. Neither A nor B

Answer A is incorrect. Technician B is also correct.

Answer B is incorrect. Technician A is also correct.

Answer C is correct. Both Technicians are correct. Many solenoids have a return spring that moves the metallic ball or rod back to the "at rest" position when the solenoid is not energized. If this spring breaks, then there is no mechanism to move the solenoid plunger or ball back to the "at rest" position. Solenoids also have a coil that can be tested with an ohmmeter. There is not a standard specification for the resistance measurement. The technician should compare the ohmmeter reading to a manufacturer's specification or to a known good solenoid.

Answer D is incorrect. Both Technicians are correct.

TASK D.7

11. The stoplights do not work when the brake pedal is depressed on a heavy-duty truck. Which of the following conditions would be the most likely cause of this problem?

   A. Faulty cruise control switch
   B. Damaged air hose at the stoplight switch
   C. Misadjusted air governor
   D. Faulty hazard flasher

Answer A is incorrect. A faulty cruise switch could cause the cruise control system to malfunction.

Answer B is correct. A damaged air hose at the stoplight switch could cause the stoplights to be inoperative due to the stop lamp switch not receiving an air signal.

Answer C is incorrect. A misadjusted air governor could cause the air pressure to not be regulated properly. The cut-out pressure should be approximately 120 psi and the cut-in pressure should be approximately 100 psi.

Answer D is incorrect. A faulty hazard flasher would cause the hazard flasher system to malfunction.

TASK A.9

12. Technician A says that a wiring diagram uses schematic symbols to represent electrical components. Technician B says that wiring diagrams usually show the splice and connector numbers for each circuit. Who is correct?

   A. A only
   B. B only
   C. Both A and B
   D. Neither A nor B

Answer A is incorrect. Technician B is also correct.

Answer B is incorrect. Technician A is also correct.

Answer C is correct. Both Technicians are correct. Wiring diagrams use symbols to represent electrical components. In addition, wiring diagrams display the splice and connector numbers that are related to the circuit.

Answer D is incorrect. Both Technicians are correct.

13. What is the function of a pyrometer?

    A. Monitors fuel temperature

    B. Indicates the battery state of charge

    C. Indicates engine speed

    D. Monitors exhaust temperature

TASK E.3

Answer A is incorrect. A fuel temperature sensor is used on some engines to monitor fuel temperature.

Answer B is incorrect. A voltmeter measures the battery state of charge, not a pyrometer.

Answer C is incorrect. A tachometer displays engine speed, not a pyrometer.

Answer D is correct. A pyrometer measures exhaust temperatures. Some manufacturers use a pyrometer system to measure and indicate exhaust temperature, which is related to engine loading. Engine loading can be due to air inlet restriction, fuel injection timing, and overfueling. In most cases, the major components of a pyrometer system are a gauge, an exhaust thermocouple, and a pyrometer control module.

14. All of the following statements about the J1939 data bus network are correct EXCEPT:

    A. The J1939 data bus is a two-wire network.

    B. The J1939 data bus can be tested at the 9-pin data connector.

    C. The J1939 data bus should have 60 ohms of resistance when tested with an ohmmeter.

    D. The J1939 data bus can communicate at 250 gigabytes per second.

TASK A.10

Answer A is incorrect. The J1939 data bus is a two-wire network that connects the modules of a truck. This communication network allows the modules to send and receive data messages at a very fast rate.

Answer B is incorrect. The J1939 data bus can be tested at the 9-pin data connector. The resistance of the network can be tested with a digital ohmmeter. A digital voltmeter or a digital oscilloscope can be used to test the voltage levels of the data bus.

Answer C is incorrect. The J1939 data bus should have approximately 60 ohms of resistance when tested with an ohmmeter. This test can be performed at the data connector and should be done only after disconnecting the truck batteries.

Answer D is correct. The J1939 data bus can currently communicate at speeds up to 250 kilobytes per second.

15. When checking open circuit battery voltage, Technician A says that a 12 volt battery is considered fully charged if a voltmeter probed across it reads anything over 12 volts. Technician B states that the charging system should produce approximately 13.5 to 14.5 volts. Who is correct?

TASK B.1

    A. A only

    B. B only

    C. Both A and B

    D. Neither A nor B

Answer A is incorrect. A battery should have 12.6 volts to be considered fully charged.

Answer B is correct. Only Technician B is correct. Normal charging system voltage is approximately 13.5 to 14.5 volts. This voltage will vary according to the electrical loads, as well as the battery charge level.

Answer C is incorrect. Only Technician B is correct.

Answer D is incorrect. Technician B is correct.

TASK B.4

16. Technician A says that battery boxes should be cleaned with mineral spirits. Technician B says that battery hold down hardware should be tight and secure to prevent the battery from moving during use. Who is correct?

    A. A only

    B. B only

    C. Both A and B

    D. Neither A nor B

    Answer A is incorrect. It would be dangerous to use mineral spirits to clean the battery box. There is a good chance that battery acid is present in the battery box, so water and baking soda should be used to clean this area.

    Answer B is correct. Only Technician B is correct. All battery braces, brackets, and hardware should be tight and secure before returning a truck to service.

    Answer C is incorrect. Only Technician B is correct.

    Answer D is incorrect. Technician B is correct.

TASK C.6

17. The alternator needs to be replaced on a heavy-duty truck. Technician A says that the negative battery cable should be removed during this process. Technician B says the drive belt should be closely inspected during this process. Who is correct?

    A. A only

    B. B only

    C. Both A and B

    D. Neither A nor B

    Answer A is incorrect. Technician B is also correct.

    Answer B is incorrect. Technician A is also correct.

    Answer C is correct. Both Technicians are correct. The negative battery cable needs to be disconnected while replacing the alternator to reduce the possibility of a short to ground. The drive belt should always be closely inspected whenever replacing the alternator.

    Answer D is incorrect. Both Technicians are correct.

TASK B.6

18. Which would be the last connection to make when connecting jumper cables to a truck with dead batteries?

    A. Positive post of the truck with the dead battery

    B. Positive post of the truck with the charged battery

    C. Negative post of the truck with the charged battery

    D. Engine block of the truck with the dead battery

    Answer A is incorrect. Connecting the engine block of the truck with the dead battery is the safest way to complete the connections when jump starting.

    Answer B is incorrect. Connecting the engine block of the truck with the dead battery is the safest way to complete the connections when jump starting.

    Answer C is incorrect. Connecting the engine block of the truck with the dead battery is the safest way to complete the connections when jump starting.

    Answer D is correct. Connecting the engine block of the truck with the dead battery is the safest way to complete the connections when jump starting.

19. How is a starter ground circuit resistance check performed?

TASK B.9,
B.10

    A. A voltmeter is connected between the ground terminal of the battery and starter ground stud and read while the engine is being cranked.

    B. An ohmmeter is connected between the starter relay housing and the starter housing.

    C. An ohmmeter is connected between the ground side of the battery and the starter housing and read while the engine is cranking.

    D. A voltmeter is connected between the positive side of the battery and the starter solenoid while the engine is off.

    Answer A is correct. A voltage drop test of the ground side of the starting circuit is performed by connecting a voltmeter between the battery ground terminal and the starter ground stud. This checks the resistance of the ground side of the circuit.

    Answer B is incorrect. An ohmmeter cannot provide enough current through the starting circuit to determine excessive resistance. This can only be done with a voltage drop test.

    Answer C is incorrect. An ohmmeter should never be connected into a live circuit. Possible meter damage may result.

    Answer D is incorrect. In order to conduct a voltage drop test, the starter must be cranking. Also, testing the positive side of the battery will not verify ground circuit performance.

20. Which of the following components is LEAST LIKELY to be part of the starter control circuit?

TASK B.10

    A. Positive battery cable

    B. Ignition switch

    C. Park/neutral switch

    D. Starter relay

    Answer A is correct. The positive battery cable is directly connected to the starter solenoid and provides power to the load (high amperage) side of the starting circuit.

    Answer B is incorrect. The ignition switch is the device that allows the driver to engage the start sequence. It is part of the control circuit.

    Answer C is incorrect. The park/neutral switch prevents the starter from engaging while in gear. It is part of the control circuit.

    Answer D is incorrect. The starter relay is in the starter control circuit. This device receives power from the ignition switch and then sends power to the control side of the solenoid.

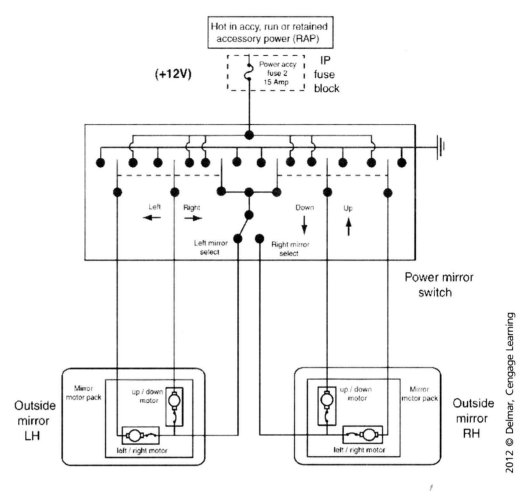

2012 © Delmar, Cengage Learning

TASK E.11

21. Referring to the figure above, the left-side mirror works correctly but the right-side mirror is totally inoperative. Which of the following conditions would most likely be the cause of this problem?

A. Faulty right-side up/down motor

B. Broken contact at the mirror select switch

C. Missing ground at the power mirror switch

D. Blown power accessory fuse

Answer A is incorrect. The left/right motor would still operate even if the up/down motor were bad.

Answer B is correct. A broken contact at the mirror select switch could cause the right power mirror to be totally inoperative.

Answer C is incorrect. A missing ground at the power mirror switch would cause all mirror operation to stop. The ground at the switch is the only ground in the power mirror circuit.

Answer D is incorrect. A blown power accessory fuse would cause all mirror operation to stop.

22. Technician A says that the replacement starter assembly should be bench tested prior to installing it on the engine. Technician B says that the starter connectors and terminals should be inspected and cleaned prior to installing a replacement starter. Who is correct?

TASK B.14

   A. A only

   B. B only

   C. Both A and B

   D. Neither A nor B

   Answer A is incorrect. Technician B is also correct.

   Answer B is incorrect. Technician A is also correct.

   Answer C is correct. Both Technicians are correct. It is a good practice to bench test the replacement starter prior to installing it into the vehicle. In addition, the starter connectors and terminals should be inspected and cleaned.

   Answer D is incorrect. Both Technicians are correct.

23. The battery housing received some damage from driving a vehicle on rough roads. Electrolyte spilled all over the battery tray. Technician A says that carburetor cleaner should be used to clean the area. Technician B says that baking soda could be used to neutralize the battery acid. Who is correct?

TASK B.4

   A. A only

   B. B only

   C. Both A and B

   D. Neither A nor B

   Answer A is incorrect. Carburetor cleaner should never be used around spilled electrolyte.

   Answer B is correct. Only Technician B is correct. Baking soda can be used to neutralize the spilled electrolyte acid of a battery.

   Answer C is incorrect. Only Technician B is correct.

   Answer D is incorrect. Technician B is correct.

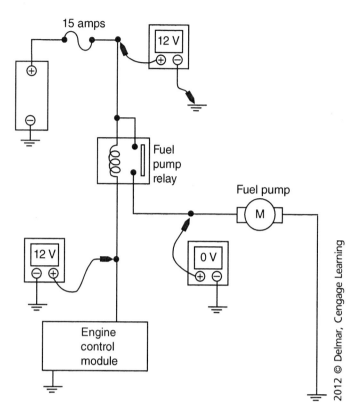

TASK A.8,
E.21

24. Referring to the figure above, the fuel pump is inoperative and the voltage readings shown were taken immediately after the circuit was closed. Which of the conditions listed below is the most likely cause?

A. Open 15 amp fuse

B. Defective engine control module (ECM)

C. Defective fuel pump relay

D. Defective fuel pump

Answer A is incorrect. The fuse is not open because the meter immediately after the fuse indicates 12 volts.

Answer B is correct. The voltmeter between the relay and the ECM indicates 12 volts. If the ECM were functioning properly, the meter would read zero volts, because this would then be the ground side of the relay (load). Because the ECM is not providing the proper ground circuit, the meter reads battery voltage because there is no current flow. Relays are often computer controlled.

Answer C is incorrect. The relay cannot be condemned because the ECM is not allowing it to be activated.

Answer D is incorrect. The fuel pump is not receiving any voltage from the relay as evidenced by the zero voltmeter reading next to the fuel pump. Therefore, the pump cannot be considered faulty.

25. Technician A says that the battery voltage can drop to 6 volts while cranking during normal conditions. Technician B says that one weak battery can cause a battery bank to perform poorly during extremely cold weather. Who is correct?

TASK B.8

A. A only
B. B only
C. Both A and B
D. Neither A nor B

Answer A is incorrect. The battery voltage level should not drop down to 6 volts during normal conditions. Weak batteries would likely cause this cranking voltage.

Answer B is correct. Only Technician B is correct. One weak battery in a battery bank can cause the truck to not perform well during extremely cold weather.

Answer C is incorrect. Only Technician B is correct.

Answer D is incorrect. Technician B is correct.

26. All of the following devices could be used as a circuit protection device EXCEPT:

TASK A.6

A. Maxi-fuse
B. Circuit breaker
C. Positive temperature coefficient (PTC) thermistor
D. Jumper wire

Answer A is incorrect. Maxi-fuses have a metallic strip that burns up when the current flow in the circuit rises above the maxi-fuse rating. Once the fuse opens, it must be replaced to have continued electrical operation.

Answer B is incorrect. Circuit breakers are made of a bi-metallic strip that bends when it gets hot. High current flow in a circuit will cause the circuit breaker to heat up, which causes the device to open its contacts.

Answer C is incorrect. Some late-model trucks use a PTC thermistor as a circuit protection device. PTC thermistors are electronic devices that have low resistance when they are at ambient temperature. The resistance of these devices increases as their temperature rises. This rising resistance limits current flow in the circuit they are protecting.

Answer D is correct. A jumper wire can be used to by-pass electrical components during the diagnostic process. Jumper wires should never be used as a circuit protection device.

27. A vehicle is being diagnosed for a charging problem. The alternator produced 125 amps during the output test and is rated at 130 amps. Technician A says that the alternator could have a bad diode. Technician B says that the alternator drive pulley could be too large in diameter. Who is correct?

TASK C.3, C.4

A. A only
B. B only
C. Both A and B
D. Neither A nor B

Answer A is incorrect. A bad diode would cause the generator to lose more of its capacity to charge.

Answer B is incorrect. A drive pulley that is too large in diameter would cause the generator to lose more of its capacity to charge.

Answer C is incorrect. Neither Technician is correct.

Answer D is correct. Neither Technician is correct. This generator is charging to nearly its total capacity during the output test. As long as the generator charges within approximately 10 percent of its rating, then there is nothing wrong.

TASK C.4

28. Technician A says that when performing an alternator output test, a voltmeter should be connected in series with the alternator output terminal and the battery ground cable. Technician B says that a carbon pile should be used when performing an alternator output test. Who is correct?

A. A only
B. B only
C. Both A and B
D. Neither A nor B

Answer A is incorrect. Voltmeters are always connected in parallel with the circuit being tested, not in series.

Answer B is correct. Only Technician B is correct. A carbon pile can draw current from the battery in excess of the alternator output. The engine needs to be running at about 1500–1600 rpm while performing this test.

Answer C is incorrect. Only Technician B is correct.

Answer D is incorrect. Technician B is correct.

TASK C.5

29. The ground side of a truck charging circuit is being tested for voltage drop. Technician A says to place the voltmeter leads on the voltage regulator ground terminal and the vehicle battery. Technician B says to place the voltmeter leads on the alternator negative terminal and the battery ground terminal. Who is correct?

A. A only
B. B only
C. Both A and B
D. Neither A nor B

Answer A is incorrect. Placing the voltmeter leads on the voltage regulator ground terminal and the vehicle battery would measure the voltage drop in the regulator ground circuit.

Answer B is correct. Only Technician B is correct. Placing the voltmeter leads on the alternator negative terminal and the battery ground terminal is the proper test procedure for measuring voltage drop on the ground side of the charging circuit. The voltage drop, in this case, should be less than 0.1 volts. If a greater voltage drop is measured, the alternator mounting might be loose or corrosion might be built up between the casing and mounting bracket.

Answer C is incorrect. Only Technician B is correct.

Answer D is incorrect. Technician B is correct.

TASK C.5

30. The charging system is being inspected during a maintenance inspection. The technician performs a voltage drop test on the charging system and finds 0.2 volts on the positive wire and finds 0.02 volts on the ground wire. Technician A says that a defective voltage regulator could cause this condition. Technician B says that defective alternator brushes could cause this condition. Who is correct?

A. A only
B. B only
C. Both A and B
D. Neither A nor B

Answer A is incorrect. A defective voltage regulator could cause a no charge condition or an overcharge condition.

Answer B is incorrect. Defective alternator brushes could cause an undercharge or a no charge condition.

Answer C is incorrect. Neither Technician is correct.

Answer D is correct. Neither Technician is correct. The voltage drop test results are within normal specifications. There is nothing wrong with this truck's charging system.

31. Technician A says that batteries can be recharged more quickly by using a low setting on the battery charger. Technician B says that batteries can be more thoroughly charged by using a high setting on the battery charger. Who is correct?

TASK B.5

   A.   A only

   B.   B only

   C.   Both A and B

   D.   Neither A nor B

Answer A is incorrect. Slow charging vehicle batteries takes more time than fast charging. However, slow charging is usually more thorough than fast charging.

Answer B is incorrect. Slow charging will more thoroughly recharge a battery than fast charging them.

Answer C is incorrect. Neither Technician is correct.

Answer D is correct. Neither Technician is correct. Fast charging vehicle batteries is quicker but slow charging vehicle batteries is more thorough.

32. A truck is being diagnosed for a charging problem. Technician A says that most charging circuits include a circuit protection device. Technician B says that all repairs should be resistant to water intrusion. Who is correct?

TASK C.7

   A.   A only

   B.   B only

   C.   Both A and B

   D.   Neither A nor B

Answer A is incorrect. Technician B is also correct.

Answer B is incorrect. Technician A is also correct.

Answer C is correct. Both Technicians are correct. The charging circuit typically includes either a fusible link or a high-rated fuse assembly. All wire repairs on the charging circuit should be performed using heat-shrink or crimp-and-seal connectors in order to keep water from entering the wires and causing corrosion.

Answer D is incorrect. Both Technicians are correct.

33. A headlight aiming procedure is being performed. Technician A says that when headlight aiming equipment is not available, headlight aiming can be checked by projecting the high beam of each light upon a screen or chart at a distance of 25 feet ahead of the headlights. Technician B says that when aiming the headlights, the vehicle should be exactly parallel to the chart or screen. Who is correct?

TASK D.2

   A.   A only

   B.   B only

   C.   Both A and B

   D.   Neither A nor B

Answer A is correct. Only Technician A is correct. When headlight aiming equipment is not available, headlight aiming can be checked by projecting the upper beam of each light upon a screen or chart at a distance of 25 feet ahead of the headlights. With the headlights on high beam, the hot (brightest) spot of each headlight should be centered on the point where the corresponding vertical and horizontal lines intersect on the screen for each headlight. The headlight adjusting screws are turned in or out to adjust the headlights vertically and/or laterally to obtain a proper aim.

Answer B is incorrect. The vehicle should be exactly perpendicular to the chart or screen.

Answer C is incorrect. Only Technician A is correct.

Answer D is incorrect. Technician A is correct.

TASK D.3

34. The headlights work on high beams but are inoperative on low beams. Technician A says that the dimmer switch could be the fault. Technician B says that both bulbs could be faulty. Who is correct?

  A. A only

  B. B only

  C. Both A and B

  D. Neither A nor B

Answer A is incorrect. Technician B is also correct.

Answer B is incorrect. Technician A is also correct.

Answer C is correct. Both Technicians are correct. A faulty dimmer switch could cause the headlights to be inoperative on low beams and possibly still work on high beams. Although unlikely, it is possible that both headlights could be blown on the low-beam filaments. A quick voltage test at the headlight assembly would produce the correct diagnosis. If the bulbs are receiving system voltage with the dimmer switch on low beams, then the bulbs would be at fault.

Answer D is incorrect. Both Technicians are correct.

TASK D.6

35. When the driver door of the vehicle is opened, the interior lights illuminate, but very dimly. Technician A says that the headlight switch may be defective. Technician B says a grounding problem to the door switch may be the problem. Who is correct?

  A. A only

  B. B only

  C. Both A and B

  D. Neither A nor B

Answer A is incorrect. The headlight switch, in normal operation, does not control interior lights.

Answer B is correct. Only Technician B is correct. A poor ground at the door switch will cause high resistance in the circuit that will cause the light to glow dimly.

Answer C is incorrect. Only Technician B is correct.

Answer D is incorrect. Technician B is correct.

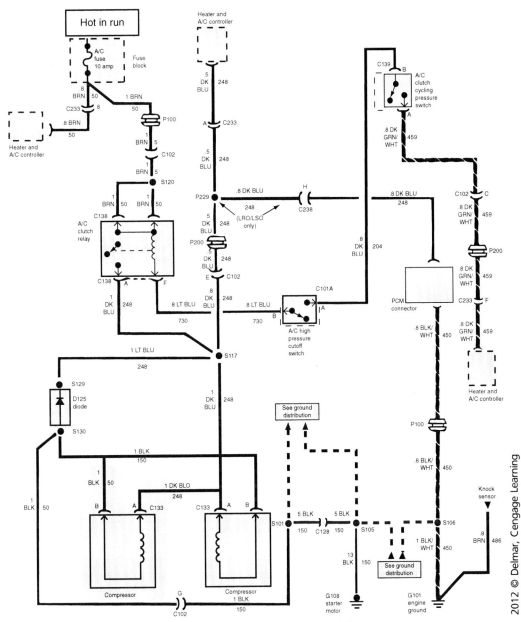

36. Referring to wiring schematic in the figure above, all of the following statements are correct EXCEPT:

TASK A.9

A. Circuit #450 uses a black wire with a white tracer.

B. Circuit #150 uses a black wire.

C. Circuit #248 uses a white wire.

D. Circuit #50 uses a brown wire.

Answer A is incorrect. Circuit #450 is the ground for the power train control module (PCM) and is colored black with a white tracer.

Answer B is incorrect. Circuit #150 leads from the diode to the compressor clutch coil and is colored black.

Answer C is correct. Circuit #248 uses a blue wire. One section of this circuit is light blue and a different section of this circuit uses a dark blue wire.

Answer D is incorrect. Circuit #50 is the supply circuit leading from the A/C fuse to the A/C clutch relay and is colored brown.

TASK D.8

37. Turn signals are being diagnosed. Technician A says that a faulty hazard flasher assembly could cause the turn signals to be inoperative. Technician B says that hazard lights and turn signals use different bulbs and sockets. Who is correct?

A.   A only

B.   B only

C.   Both A and B

D.   Neither A nor B

Answer A is incorrect. A faulty hazard flasher would not typically cause a problem with the turn signals.

Answer B is incorrect. The turn signals and hazard signals use the same bulbs and sockets.

Answer C is incorrect. Neither Technician is correct.

Answer D is correct. Neither Technician is correct. The hazard flasher is typically separate from the turn signal flasher and could not cause this problem. The same bulbs and sockets are typically used for the hazard lamps and the turn signal lamps.

TASK D.9

38. All of the following statements are correct concerning turn signal operation on a heavy-duty truck EXCEPT:

A.   Some turn signal bulbs share a filament with the stoplights.

B.   Some turn signal bulbs use the same socket as the back-up lights.

C.   The turn signal flasher is sometimes combined with the hazard flasher.

D.   Some turn signal systems incorporate a relay to make the turn signals flash correctly.

Answer A is incorrect. Many truck lighting systems are designed to allow the turn signals and the stoplights to share a bulb.

Answer B is correct. Turn signals do not use the same socket as the back-up lights. These lights have to function independently of each other.

Answer C is incorrect. The turn signal and hazard flasher can be combined into one unit on some heavy-duty trucks.

Answer D is incorrect. Some late-model heavy-duty trucks use a relay to flash the turn signals and hazard lights.

TASK E.1

39. All the gauges are erratic in an instrument panel with thermal-electric gauges and an instrument voltage limiter. Technician A says the alternator may be at fault. Technician B says the instrument voltage limiter may be defective. Who is correct?

A.   A only

B.   B only

C.   Both A and B

D.   Neither A nor B

Answer A is incorrect. The instrument voltage limiter maintains a constant voltage value to the gauges regardless of battery voltage.

Answer B is correct. Only Technician B is correct. A defective instrument voltage limiter will affect all the dash gauges. Some gauges were designed to operate under very specific, constant voltages in order to provide an accurate reading. To protect these gauges against heavy voltage fluctuations that can occur due to the charging system operation, for example, some systems require the use of an instrument voltage regulator (IVR). A typical IVR consists of a set of normally closed contacts and a heating coil wrapped around a bi-metallic strip. Another name for the IVR is a voltage limiter because it can limit the voltage to the gauge circuit to 5 volts, regardless of the charging system voltage.

Answer C is incorrect. Only Technician B is correct.

Answer D is incorrect. Technician B is correct.

40. All of the following conditions could cause the temperature gauge to read "high" EXCEPT:

    A. Shorted temperature sending unit

    B. Grounded temperature sending unit wire

    C. Open instrument panel fuse

    D. Faulty instrument panel circuit board

TASK E.1,
E.12

Answer A is incorrect. A shorted temperature sending unit could cause the temperature gauge to read high due to increased current flow in the sending unit circuit.

Answer B is incorrect. A grounded temperature sending unit wire could cause the temperature gauge to read high due to increased current flow in the sending unit circuit.

Answer C is correct. An open instrument panel fuse would cause all of the gauges to be inoperative.

Answer D is incorrect. A faulty instrument panel circuit board could cause the temperature gauge to read high.

41. Technician A says that a high-impedance digital meter is needed to perform voltage tests on the data bus network. Technician B says that an oscilloscope can be used to view the communication activity on the data bus network. Who is correct?

    A. A only

    B. B only

    C. Both A and B

    D. Neither A nor B

TASK A.10

Answer A is incorrect. Technician B is also correct.

Answer B is incorrect. Technician A is also correct.

Answer C is correct. Both Technicians are correct. Testing on the data bus network can be done with high-impedance digital meters. In addition, an oscilloscope can be used to view the data bus wires for electrical activity.

Answer D is incorrect. Both Technicians are correct.

42. Which of the following tools could be used to test a coolant temperature sending unit?

    A. Terminal removal tool

    B. Jumper wire

    C. Digital multimeter (DMM)

    D. Test light

TASK E.3

Answer A is incorrect. A terminal removal tool is a good choice for use when removing or replacing terminals on heavy-duty trucks. Care should be taken not to damage the plastic connector body when removing terminals.

Answer B is incorrect. A jumper wire can be used to by-pass switches and relays during the diagnostic process.

Answer C is correct. A DMM could be used to measure the resistance of the coolant temperature sending unit. The resistance of the sending unit should vary as the temperature changes.

Answer D is incorrect. A test light can be used to test fuses and other electrical components for the presence of voltage. The test light is not accurate enough to closely measure voltage levels. A digital voltmeter would be needed to accurately measure voltage.

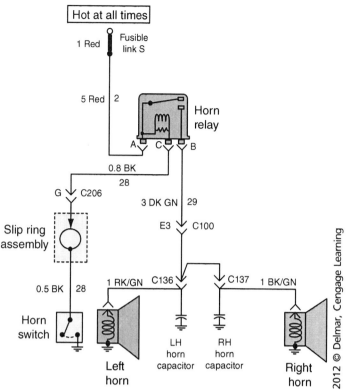

TASK E.6

43. Referring to the figure above, the problem is a horn that sounds continuously. Technician A says that a short to ground at connector C206 may cause this. Technician B states that a short to ground at connector C100 could cause this problem. Who is correct?

A. A only
B. B only
C. Both A and B
D. Neither A nor B

Answer A is correct. Only Technician A is correct. Connector C206 is located on the switch side of the relay coil. If the circuit was shorted to ground at this location, the relay would operate all of the time, which would cause the horn to blow continuously.

Answer B is incorrect. A short to ground at C100 would cause a blown fusible link S due to excessive current flow when the horn switch is depressed.

Answer C is incorrect. Only Technician A is correct.

Answer D is incorrect. Technician A is correct.

TASK E.8

44. A wiper motor fails to operate. Which of the following would be the LEAST LIKELY cause?

A. Defective wiper switch
B. Tripped thermal overload protector
C. Tripped circuit breaker
D. High resistance in the motor wiring

Answer A is incorrect. A defective wiper switch would interrupt current flow to the wiper motor and would not allow motor operation.

Answer B is incorrect. A tripped thermal overload protector would open the circuit and prevent current flow to the motor.

Answer C is incorrect. A tripped circuit breaker would open the circuit and not allow current flow to the motor.

Answer D is correct. High resistance in the wiring should only slow the motor down, not stop it.

45. The starter solenoid performs all of the following functions EXCEPT:

    A. Provides a path for high current to flow into the starter
    B. Pushes the drive gear out to the flywheel
    C. Connects the "bat" terminal to the "motor" terminal
    D. Provides gear reduction to increase torque in the starter

TASK B.12

Answer A is incorrect. The starter solenoid provides a path for high current to flow into the starter.

Answer B is incorrect. The starter solenoid provides the linear movement to push the drive gear into the flywheel.

Answer C is incorrect. The starter solenoid connects the "bat" terminal to the "motor" terminal when it is energized.

Answer D is correct. The starter solenoid does not provide any gear reduction for the starter.

46. What is the most likely cause for a windshield wiper system that only works on low speed?

    A. Blown fuse
    B. Faulty multi-function switch
    C. Loose ground at the wiper motor
    D. Open park switch

TASK E.9

Answer A is incorrect. A blown fuse would cause the wipers to be totally inoperative at all speeds.

Answer B is correct. The multi-function switch often contains the wiper switch in addition to several other switches, such as the turn signal switch, the hazard switch, the headlight switch, and sometimes the cruise control switch.

Answer C is incorrect. A loose wiper motor ground would cause all of the speeds to be negatively affected.

Answer D is incorrect. An open park switch on the windshield wiper motor would cause the wipers to fail to return to the rest/park position when the switch is turned off.

47. All of the following wiper problems could cause slow wiper operation EXCEPT:

    A. Weak batteries
    B. Poor alternator output
    C. Worn wiper blades
    D. Excessively tight wiper linkage

TASK E.9

Answer A is incorrect. Very weak truck batteries could cause slow wiper operation during times when the truck is idling with several electrical loads running.

Answer B is incorrect. Poor alternator output could cause slow wiper operation because of reduced amperage being supplied to the truck batteries.

Answer C is correct. Worn wiper blades can cause poor windshield clearing, but would not cause slow wiper operation.

Answer D is incorrect. Tight wiper linkage could cause slow wiper operation due to the increased physical resistance to motion.

TASK E.10

48. A truck has a windshield washer pump system that does not operate properly. Which of the following conditions would be the most likely cause of this problem?

A. Binding wiper linkage
B. Faulty wiper park switch
C. Faulty windshield wiper motor
D. Clogged washer hose

Answer A is incorrect. Binding wiper linkage could cause the windshield wipers to operate slowly or possibly not at all.

Answer B is incorrect. A faulty wiper park switch could cause the windshield wipers to not park correctly when turned off.

Answer C is incorrect. A faulty windshield wiper motor would cause the windshield wipers to not function.

Answer D is correct. A clogged washer hose could cause the washer pump system to not operate correctly. If this were the problem, then the technician would hear the washer pump running without sending washer solvent to the windshield.

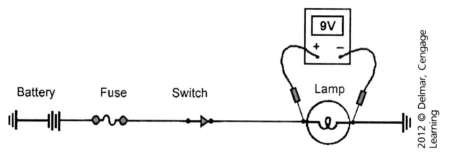

TASK A.1

49. Referring to the figure above, the 12 volt battery is fully charged and the switch is closed. Which of the following conditions would be the most likely cause of this measurement?

A. Wire repair that used a larger than specified wire
B. Faulty bulb
C. Blown fuse
D. Loose terminal connection at the switch

Answer A is incorrect. A larger wire used in a wire repair would not cause a voltage loss.

Answer B is incorrect. There is no evidence that the bulb is faulty. The meter shows only 9 volts being dropped across the bulb.

Answer C is incorrect. A blown fuse would prevent any voltage from getting past the fuse contacts.

Answer D is correct. A loose terminal connection at the switch could cause a voltage loss at the switch. Performing a voltage drop test on the switch is a good way to isolate the problem.

TASK E.13

50. The auxiliary power outlet is inoperative and the fuse is found to be open. What is the LEAST LIKELY cause for this condition?

A. Shorted power wire near the auxiliary connector
B. Foreign metal object in the auxiliary power outlet
C. Faulty electrical device connected to the outlet
D. Open internal connection at the power outlet

Answer A is incorrect. A shorted power wire could cause the fuse for the auxiliary power outlet to blow.

Answer B is incorrect. A foreign metal object in the auxiliary power outlet could cause the fuse to blow.

Answer C is incorrect. A faulty electrical device that is connected to the auxiliary power outlet could cause the fuse to blow.

Answer D is correct. An open circuit will not cause a fuse to blow.

Appendices

PREPARATION EXAM ANSWER SHEET FORMS

ANSWER SHEET

1. _____	21. _____	41. _____
2. _____	22. _____	42. _____
3. _____	23. _____	43. _____
4. _____	24. _____	44. _____
5. _____	25. _____	45. _____
6. _____	26. _____	46. _____
7. _____	27. _____	47. _____
8. _____	28. _____	48. _____
9. _____	29. _____	49. _____
10. _____	30. _____	50. _____
11. _____	31. _____	
12. _____	32. _____	
13. _____	33. _____	
14. _____	34. _____	
15. _____	35. _____	
16. _____	36. _____	
17. _____	37. _____	
18. _____	38. _____	
19. _____	39. _____	
20. _____	40. _____	

ANSWER SHEET

1. _____   21. _____   41. _____
2. _____   22. _____   42. _____
3. _____   23. _____   43. _____
4. _____   24. _____   44. _____
5. _____   25. _____   45. _____
6. _____   26. _____   46. _____
7. _____   27. _____   47. _____
8. _____   28. _____   48. _____
9. _____   29. _____   49. _____
10. _____   30. _____   50. _____
11. _____   31. _____
12. _____   32. _____
13. _____   33. _____
14. _____   34. _____
15. _____   35. _____
16. _____   36. _____
17. _____   37. _____
18. _____   38. _____
19. _____   39. _____
20. _____   40. _____

ANSWER SHEET

1. _____	21. _____	41. _____
2. _____	22. _____	42. _____
3. _____	23. _____	43. _____
4. _____	24. _____	44. _____
5. _____	25. _____	45. _____
6. _____	26. _____	46. _____
7. _____	27. _____	47. _____
8. _____	28. _____	48. _____
9. _____	29. _____	49. _____
10. _____	30. _____	50. _____
11. _____	31. _____	
12. _____	32. _____	
13. _____	33. _____	
14. _____	34. _____	
15. _____	35. _____	
16. _____	36. _____	
17. _____	37. _____	
18. _____	38. _____	
19. _____	39. _____	
20. _____	40. _____	

ANSWER SHEET

1. _____	21. _____	41. _____
2. _____	22. _____	42. _____
3. _____	23. _____	43. _____
4. _____	24. _____	44. _____
5. _____	25. _____	45. _____
6. _____	26. _____	46. _____
7. _____	27. _____	47. _____
8. _____	28. _____	48. _____
9. _____	29. _____	49. _____
10. _____	30. _____	50. _____
11. _____	31. _____	
12. _____	32. _____	
13. _____	33. _____	
14. _____	34. _____	
15. _____	35. _____	
16. _____	36. _____	
17. _____	37. _____	
18. _____	38. _____	
19. _____	39. _____	
20. _____	40. _____	

ANSWER SHEET

1. _____	21. _____	41. _____
2. _____	22. _____	42. _____
3. _____	23. _____	43. _____
4. _____	24. _____	44. _____
5. _____	25. _____	45. _____
6. _____	26. _____	46. _____
7. _____	27. _____	47. _____
8. _____	28. _____	48. _____
9. _____	29. _____	49. _____
10. _____	30. _____	50. _____
11. _____	31. _____	
12. _____	32. _____	
13. _____	33. _____	
14. _____	34. _____	
15. _____	35. _____	
16. _____	36. _____	
17. _____	37. _____	
18. _____	38. _____	
19. _____	39. _____	
20. _____	40. _____	

ANSWER SHEET

1. _____	21. _____	41. _____
2. _____	22. _____	42. _____
3. _____	23. _____	43. _____
4. _____	24. _____	44. _____
5. _____	25. _____	45. _____
6. _____	26. _____	46. _____
7. _____	27. _____	47. _____
8. _____	28. _____	48. _____
9. _____	29. _____	49. _____
10. _____	30. _____	50. _____
11. _____	31. _____	
12. _____	32. _____	
13. _____	33. _____	
14. _____	34. _____	
15. _____	35. _____	
16. _____	36. _____	
17. _____	37. _____	
18. _____	38. _____	
19. _____	39. _____	
20. _____	40. _____	

Glossary

Actuator A device that delivers motion in response to an electrical signal.

Alternator A device that converts mechanical energy from the engine to electrical energy used to charge the battery and power various vehicle accessories.

Ammeter A device (usually part of a digital multimeter or DMM) that is used to measure current flow in units known as amps or milliamps.

Ampere A unit for measuring electrical current, also known as amp.

Ampere Hours (AH) An older method of determining a battery's capacity.

Analog Signal A voltage signal that varies within a given range from high to low, including all points in between.

Analog-to-Digital Converter (A/D Converter) A device that converts analog voltage signals to a digital format, located in the section of a control module called the input signal conditioner.

Analog Volt/Ohmmeter (AVOM) A test meter used for checking voltage and resistance. These are older-style meters that use a needle to indicate the values being read. They should not be used with electronic circuits.

Armature The rotating component of a (1) starter or other motor, (2) generator, (3) compressor clutch.

ATA Connector American Trucking Association data link connector. The standard connector used by most manufacturers for accessing data information from various electronic systems in trucks.

Auto-Ranging DMM A digital multimeter that automatically adjusts the scale for the circuit being tested.

Blade Fuse A type of fuse having two flat male lugs for insertion into mating female sockets.

Blower Fan A fan that pushes air through a ventilation, heater, or air conditioning (HVAC) system.

Cartridge Fuse A type of fuse having a strip of low melting point metal enclosed in a glass tube.

CCA Acronym for cold cranking amps, a common method used to specify battery capacity.

CCM Acronym for chassis control module, a computer used to control various aspects of driveline operation. Usually does not include any engine controls.

CEL Acronym for check engine light.

Circuit A complete path for electrical current to flow.

Circuit Breaker A circuit protection device used to open a circuit when current in excess of its rated capacity flows through a circuit. Designed to reset, either manually or automatically.

Data Bus Data backbone of the chassis electronic system using hardware and communications protocols consistent with CAN 2.0 and SAE J-1939 standards.

Data Link A dedicated wiring circuit in the system of a vehicle used to transfer information from one or more electronic systems to a diagnostic tool, or from one module to another.

Diode An electrical one-way check valve. It allows current flow in one direction but not the other.

DLC Acronym for data link connector.

DMM Acronym for digital multimeter, a tool used for measuring circuit values such as voltage, current flow, and resistance. The meter has a digital readout and is recommended for measuring sensitive electronic circuits.

DTC Acronym for diagnostic trouble code. A DTC is a piece of troubleshooting data that is recorded when an on-board computer senses a problem.

ECM/ECU Acronyms for electronic control module/electronic control unit, the modules that control the electronic systems on a truck.

Electricity The flow of electrons through various circuits, usually controlled by manual switches and senders.

Electronically Erasable Programmable Memory (EEPROM) Computer memory that enables write-to-self, logging of failure codes and strategies, and customer/proprietary data programming.

Electronics The branch of electricity in which electrical circuits are monitored and controlled by a computer, the purpose of which is to allow for more efficient operation of those systems.

Electrons Negatively charged particles orbiting every atomic nucleus.

EMI Acronym for electromagnetic interference, low-level radiation that interferes with electrical/electronically controlled circuits, causing erratic outcomes.

Fault Code A code stored in computer memory to be retrieved by a technician using a diagnostic tool.

Full-Fielding A process that by-passes the voltage regulator on a charging system. This test should only be performed for a few seconds on systems due to the potential to cause damage to the electronic components on the vehicle.

Fuse A circuit protection device designed to open a circuit when amperage that exceeds its rating flows through a circuit.

Fusible Link A short piece of wire with a special insulation designed to melt and open during an overload. Installed near the power source in a vehicle to protect one or more circuits, it is usually two to four wire gauge sizes smaller than the circuit it is designed to protect.

Grounded Circuit A condition that causes current to return to the battery before reaching its intended destination. Because the resistance is usually much lower than normal, excess current flows and damage to wiring or other components usually result. Also known as a short circuit.

Halogen Light A lamp having a small quartz/glass bulb that contains a filament surrounded by halogen gas. It is contained within a larger metal reflector and lens element.

Harness and Harness Connectors The routing of wires along with termination points to allow for vehicle electrical operation.

High-Resistance Circuits Circuits that have resistance in excess of what was intended. Causes a decrease in current flow along with dimmer lights and slower motors.

Impedance Internal resistance of a digital meter.

In-Line Fuse An electrical protection device that is usually mounted in a special holder inserted somewhere into a circuit and near a power source.

Insulator A material, such as rubber or glass, that offers high resistance to the flow of electricity.

Integrated Circuit A solid-state component containing diodes, transistors, resistors, capacitors, and other electronic components mounted on a single piece of material and capable of performing numerous functions.

ISO Acronym for International Organization for Standardization.

IVR Acronym for instrument voltage regulator, a device that regulates the voltage going to various dash gauges to a certain level to prevent inaccurate readings. Usually used with bi-metal type gauges.

J1939 Network A data circuit used on late-model cars and trucks that allows several computers to share information.

Jump Start A term used to describe the procedure in which a booster battery is used to help start a vehicle with a low or dead battery.

Jumper Wire A piece of test wire, usually with alligator clips on each end, meant to by-pass sections of a circuit for testing and troubleshooting purposes.

LVD Acronym for low voltage disconnect.

Magnetic Switch The term usually used to describe a relay that switches power from the battery to a starter solenoid. It is controlled by the start switch.

Maintenance-Free Battery A battery that does not require the addition of water during its normal service life.

Maxi-fuse A fuse that is used to protect several electrical circuits. These components took the place of fusible links.

Milliamp 0.001 amp; 1000 milliamps = 1 amp.

Millivolt 0.001 volt; 1000 millivolts = 1 volt.

NC Acronym for normally closed.

Negative Temperature Coefficient (NTC) Thermistor A thermistor that operates by decreasing its resistance as the temperature increases.

NO Acronym for normally open.

Ohm A unit of electrical resistance.

Ohmmeter An instrument used to measure resistance in an electrical circuit, usually part of a digital multimeter (DMM).

Ohm's Law A basic law of electricity stating that, in any electrical circuit, voltage, amperage, and resistance work together in a mathematical relationship.

OL Acronym for out of limits.

Open Circuit A circuit in which current has ceased to flow because of either an accidental breakage (such as a broken wire) or an intentional breakage (such as opening a switch).

Output Driver An electronic on/off switch that a computer uses to drive higher amperage outputs, such as injector solenoids.

Parallel Circuit An electrical circuit that provides two or more paths for the current to flow. Each path has separate resistances (or loads) and operates independently from the other parallel paths. In a parallel circuit, amperage can flow through more than one load path at a time.

PCM Acronym for power train control module. A computer that operates the engine and functions on an electronic engine.

Piezo Resistor A device that creates a voltage when exposed to something that produces pressure or creates a vibration.

Positive Temperature Coefficient (PTC) Thermistor A thermistor that operates by increasing its resistance as the temperature increases.

Power A measure of work being done. In electrical systems, this is measured in watts, which is simply amps × volts.

Processor The brain of the processing cycle in a computer or module. Performs data fetch-and-carry, data organization, logic, and arithmetic computation.

Programmable Read Only Memory (PROM) An electronic memory component that contains program information specific to chassis applications; used to qualify ROM data. PROMs are used in most of the computers on cars and trucks.

PTC Acronym for positive temperature coefficient.

Random Access Memory (RAM) The memory used during computer operation to store temporary information. The computer can write, read, and erase information from RAM in any order, which is why it is called random. RAM is electronically retained and therefore volatile.

Read Only Memory (ROM) A type of memory used in computers to store information permanently.

Reference Voltage The voltage supplied to various sensors by the computer, which acts as a baseline voltage; modified by sensors to act as input signal.

Relay An electrical switch that uses a small current to control a larger one, such as a magnetic switch used in starter motor cranking circuits.

Reserve Capacity Rating The measurement of the ability of a battery to sustain a minimum vehicle electrical load in the event of a charging system failure.

Resistance The opposition to current flow in an electrical circuit; measured in units known as ohms.

RFI Acronym for radio frequency interference. A disturbance caused by electromagnetic induction that can cause computer problems on a car or truck.

Rotor (1) A part of the alternator that provides the magnetic fields necessary to generate a current flow. (2) The rotating member of an assembly.

Semiconductor A solid-state device that can function as either a conductor or an insulator, depending on how its crystalline structure is arranged.

Sensing Voltage A reference voltage put out by the alternator that allows the regulator to sense and adjust charging system output voltage.

Sensor An electrical unit used to monitor conditions in a specific circuit to report back to another device, such as a computer, light, solenoid, etc.

Series Circuit A circuit that consists of one or more resistances connected to a voltage source so there is only one path for electrons to flow.

Series/Parallel Circuit A circuit designed so that both series and parallel combinations exist within the same circuit.

Shore Power An electrical and communication system that is available at some truck stops across the United States to which truck drivers can connect their trucks. This connection provides electricity and communication outlets to the truck.

Short Circuit A condition, usually undesirable, between one circuit relative to ground or one circuit relative to another. Commonly caused by two wires rubbing together and exposing bare wires. It almost always causes blown fuses and/or undesirable responses.

Signal Generators Electromagnetic devices used to count pulses produced by a reluctor or chopper wheel (such as teeth on a transmission output shaft gear), which are then translated by an engine control module (ECM) or gauge to display speed, RPM, etc.

Slip Rings and Brushes Components of an alternator that conduct current to the rotating rotor. Most alternators have two slip rings mounted directly on the rotor shaft that are insulated from the shaft and each other. A spring-loaded carbon brush is located on each slip ring to carry the current to and from the rotor windings.

Solenoid An electromagnet used to perform mechanical work, made with one or two coil windings wrapped around an iron tube. A good example is a starter solenoid, which shifts the starter drive pinion into mesh with the flywheel ring gear.

Spoon An electrical tool that can be used to test live voltage readings. A spoon is another name for a backprobing tool.

Starter (Neutral) Safety Switch A switch used to insure that a starter is not engaged when the transmission is in gear.

Switch A device used to control current flow in a circuit. It can be either manually operated or controlled by another source, such as a computer.

Test Light An electrical test tool that uses a light bulb to indicate the presence of voltage. This tool should not be used on electronic circuits due to the potential to cause damage.

Thermistor A variable resistor that changes its resistance with temperature changes. Thermistors can be negative temperature coefficient (NTC) or positive temperature coefficient (PTC).

Transistor An electronic device that acts as a switching mechanism.

Volt A unit of electrical force, or pressure.

Voltage Drop The amount of voltage lost in any particular circuit due to excessive resistance in one or more wires, conductors, etc., either leading up to or exiting from a load (e.g., starter motor). Voltage drops can only be checked when the circuit is energized.

Voltmeter A device, usually incorporated into a DMM, used to measure voltage.

Watt A unit of electrical power, calculated by multiplying volts × amps.

Windings (1) The three separate bundles in which wires are grouped in an alternator stator. (2) The coil of wire found in a relay or other similar device. (3) The part of an electrical clutch that provides a magnetic field.

Xenon Headlights High-voltage, high-intensity headlamps that use heavy xenon gas elements.

Zener Diode A diode that can be reverse biased without causing damage.

Notes

Notes

Notes

Notes

Notes

Notes

Notes

CPSIA information can be obtained
at www.ICGtesting.com
Printed in the USA
FFOW02n2355140714
6349FF

9 781111 129026